I0605280

THE ULTIMATE
WINE LOVER'S TRAVEL GUIDE

First published in the UK by Sona Books, an imprint of Danann Media Publishing Limited

CAT NO. SONO579

Contributors: Jane Anson, Amanda Barnes, Filippo Bartolotta, Ceil Miller Bouchet, Carolyn Boyd, Carla Capalbo, Danielle Costley, Vicki Denig, Jennifer Dombrowski, Jessica Dupuy, Sarah Jane Evans, Helen Farrell, Harry Fawkes, Marisa Finetti, Damien Gabet, Rosemary George, Tim Hall, Anthony Hanson, Shawn Hennessey, Brooke Herron, Panos Kakaviatos, Katie Kelly Bell, Jess Lander, Sarah Lane, Angela Lloyd, Nicole MacKay, Ewan MacCormick, Bridget McGrouther, Chris Mercer, Alicia Miller, Sorrel Moseley-Williams, Adrian Mourby, Lauren Mowery, Wendy Narby, Helena Nicklin, Yolanda Ortiz de Arri, Alessandra Piubello, Ines Salpico, Fiona Sims, Aidy Smith, Darren Smith, Tyson Stelzer, Oliver Styles, Monty Waldin and Matt Walls.

Cover design: Darren Grice
Book design: Kate Cerpnjak
Editor: Martin Corteel
Proof readers: Juliette O'Neill and Cameron Thurlow

ISBN: 978-1-915343-41-3

in association with

THE ULTIMATE WINE LOVER'S TRAVEL GUIDE

Capezzana barrel cellar (see page 223)

With a foreword by **Amy Wislocki**
Editor of *Decanter* magazine

CONTENTS

FOREWORD

Let us transport you, armchair travellers, to a world of wine destinations just waiting to be explored. Wine tourism is booming, and at *Decanter* it has been our pleasure to guide readers through the most famous regions around the globe – and along the wine roads less travelled, both in our magazine pages and online.

There has never been a better time to get among the vines, visit your favourite wine estates and taste the latest vintage in its place of origin. In every wine region, young and old, Old World and New World, hardworking producers have realised that there is no better way to gain loyal custom than pouring their wine for visitors, after a tour of the vineyard and cellars.

And what better way to really understand a wine, to enhance your appreciation when you next swirl it in the glass to release the aromas, and take a sip? *Decanter's* long-time columnist Andrew Jefford urged readers recently to 'Go there. Look around. Stand in the landscape.' Meet the growers behind the wines, feel the vineyard soils under your feet, look around you and understand the topography and the terroirs that give the wine its character. And then enjoy the wines alongside the regional cuisine that invariably provides the perfect food pairing.

Happily, the classic regions where wine is made are often also among the world's most breathtakingly beautiful landscapes, guaranteed to enchant every visitor, wine lover or not. Whether

Sunset landscape in Bordeaux, France

it's the timbered buildings and cobblestone streets of Alsace's historic villages, the gently rolling, vine-carpeted hills of Piedmont in northern Italy, or the dramatic, mountainous backdrop of Argentina's Mendoza region – home to inky Malbecs and juicy steaks – these are landscapes to fall in love with, memories just waiting to be made.

In the pages that follow, we jump from continent to continent, from classic regions to emerging wine hotspots, from buzzing cities and coastal towns, to vine-covered valley floors and vineyards planted at dizzying heights. Our expert writers know these regions and their wines inside out, and on these pages share their inside track on the wineries to visit, the hotels and restaurants to book into, and what's new and exciting in the area. Everything you need to plan the perfect wine holiday. So settle back, pour a glass of wine, and enjoy the journey.

Amy Wislocki

Editor, *Decanter* magazine

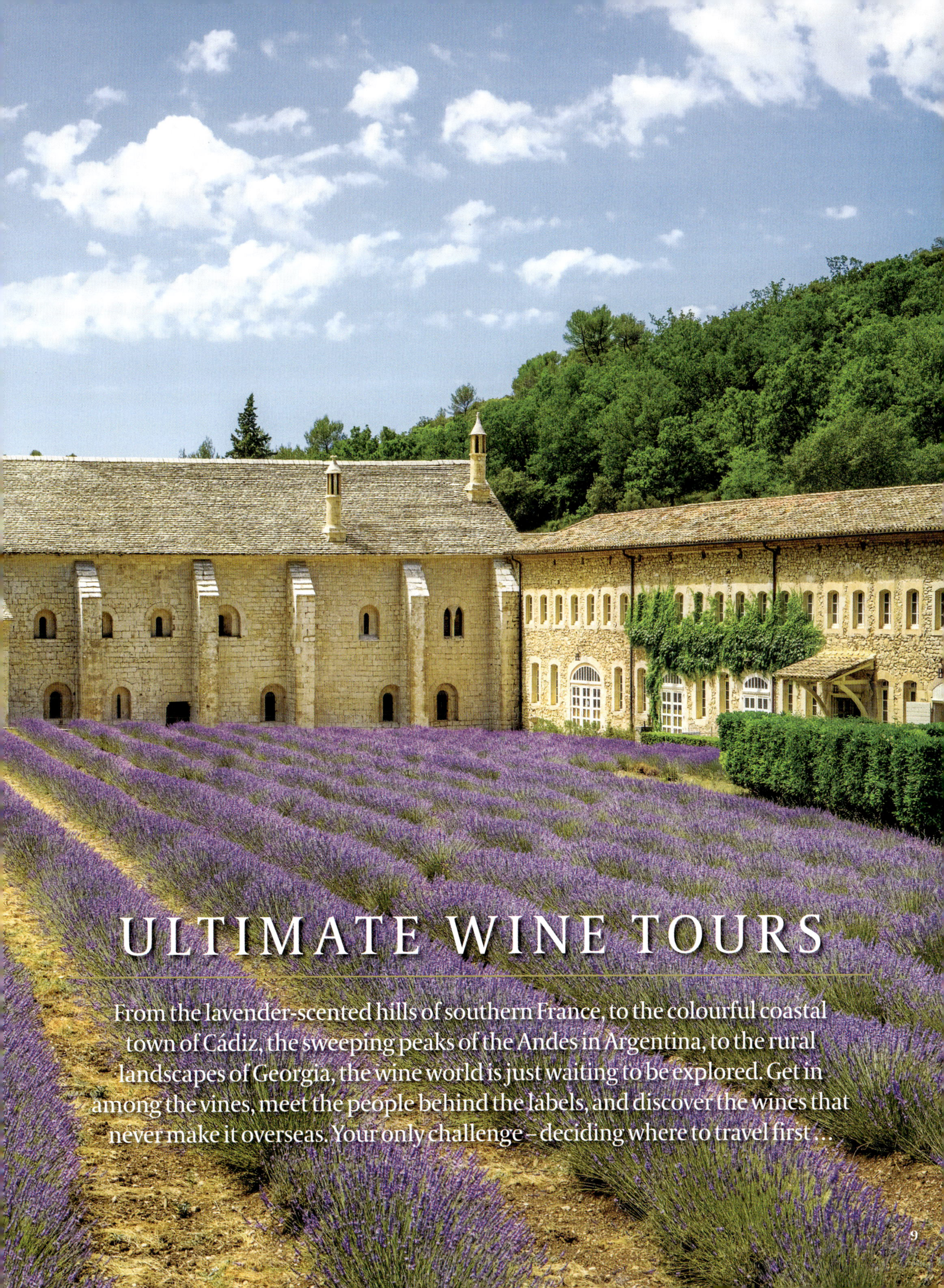

ULTIMATE WINE TOURS

From the lavender-scented hills of southern France, to the colourful coastal town of Cádiz, the sweeping peaks of the Andes in Argentina, to the rural landscapes of Georgia, the wine world is just waiting to be explored. Get in among the vines, meet the people behind the labels, and discover the wines that never make it overseas. Your only challenge – deciding where to travel first …

FRANCE

The first stop for many wine lovers planning a holiday, and little wonder, with so many famous and classic wine styles to offer. In years past, Bordeaux was famous for its aloofness, grand châteaux gates closed to visitors. Today estates are queuing up to welcome tourists. And it's the same across the whole country – Alsace, Loire, Rhône, Champagne and beyond …

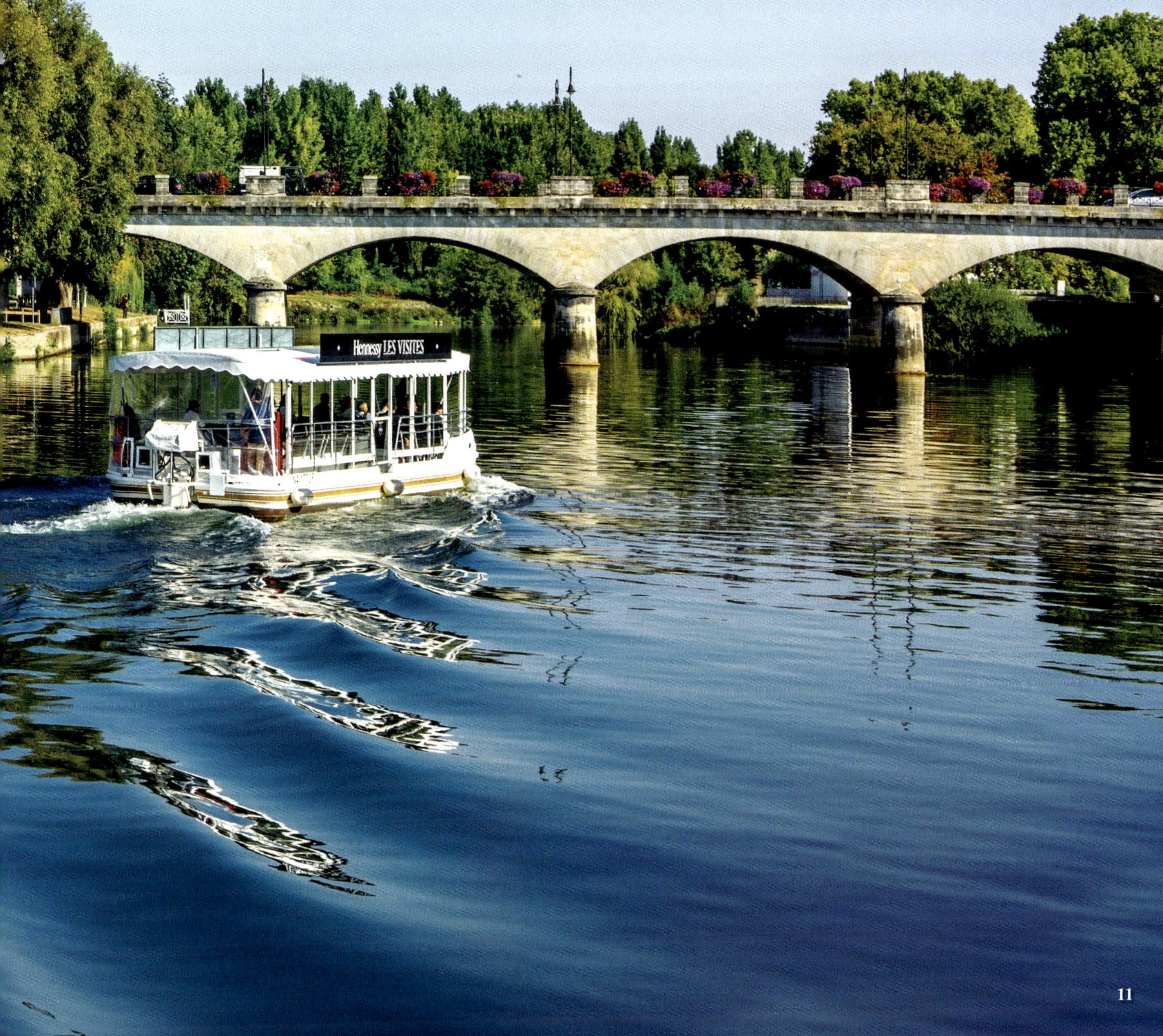

ALSACE

Perched on the French-German border, Strasbourg's impressive architecture reflects centuries of bridging two cultures. Drive out of the city and you're straight into the beautiful Alsace vineyards, with 51 grands crus along your route.

Traditional timber-framed houses in Strasbourg's La Petite France quarter

The cultural history of Strasbourg – one of France's largest cities, and the capital of Alsace – is reflected in its mix of architectural styles. Medieval timber-framed dwellings in the touristy 'Petite France' quarter (see p83) denote a thousand-year link to the Germanic Holy Roman Empire, as does the Strasbourg cathedral: built between 1015 and 1439, it is considered an exceptional example of Gothic architecture. After the Thirty Years' War, Louis XIV made Alsace part of the French kingdom, only for it to later change hands twice with Germany in the 19th and 20th centuries, which explains some impressive Wilhelmine buildings at Place de la République.

Such architectural variety matches a hodge-podge of the region's vineyard soils: from sandstone and marl to limestone and granite. Located in northeastern France, Alsace is one of France's driest and sunniest wine regions; its cool summer nights and warm days permit the grapes to ripen evenly.

Vineyard visits

A 20-minute drive from Strasbourg, and you're in wine country. Many of the quality producers are family-owned wineries, opening doors for wine tastings in their home estates, but be sure to schedule an appointment beforehand.

Old wooden casks form part of the scenery at Domaine Frédéric Mochel (*mochel.alsace*) in the village of Traenheim, west of Strasbourg, where 14 generations of the family have lived since 1669. In my most recent visit this year, I met New York importer Bradley Cohen, who fell in love with the wines during a Zoom tasting in 2021. Indeed, top Alsace restaurants feature Domaine Frédéric Mochel wines, and do not miss the bold and elegant dry Altenberg de Bergbieten Grand Cru Riesling. An expansive tasting room includes comfortable seating where the Mochel family offers bread and cheese to go with the wines.

A five-minute drive east from there, Mélanie Pfister of the eponymous estate in Dahlenheim (*melaniepfister.fr*) counts among the most talented Alsace winemakers. The eighth

generation of her family to run the estate, she crafts dry wines – from superb sparklers to pristine Pinot Gris – which also are featured in top Alsace restaurants. Try her marvellous Engelberg Grand Cru Riesling in the tasting room that once was the 18th-century vinification cellar. If the weather is sunny, find time for a 20-minute walk from town to the 300m-high south-facing Engelberg vineyard, where you not only see vines on the limestone/clay terroir, but also appreciate a great view of Strasbourg's cathedral.

Heading south

An hour's drive from Dahlenheim is the southern Alsace village of Hunawihr. It was thanks to Romain Iltis, sommelier at two-star Michelin Villa René Lalique (see below), that I discovered Domaine Mader's Rosacker Grand Cru Riesling *(vins-mader.com)*: the wine was so clean, crisp and pure that a week later I visited the estate, where co-owner Jérôme Mader emphasises wet stone minerality. Every single wine that I tried at the family winery was delicious, but Rosacker has the coolest limestone terroir in Alsace: favourable for climate change, Mader explained.

Drive another 15 minutes further south to the village of Niedermorschwihr to visit one of the most celebrated estates in Alsace these days. At Domaine Albert Boxler

YOUR ALSACE ADDRESS BOOK

ACCOMMODATION

Hôtel Cour du Corbeau, Strasbourg
Very close to the cathedral in a setting that dates back to the 16th century, this four- star hotel combines Middle Ages charm with modern amenities. Visiting friends give rave reviews. **cour-corbeau.com**

Hôtel Léonor, Strasbourg
Opened last year, enjoy casually chic (and comfortable) rooms in what was a central police station in Strasbourg. The elegant façade is a historical monument. Dine at the hotel restaurant, managed by the team from the exceptional two-star Michelin La Fourchette des Ducs. **leonor-hotel.com**

Les Haras, Strasbourg
Conveniently close to the Petite France quarter, it used to be the 18th-century National Stud Farm and Equestrian Academy, but now impresses with stylish rooms and gorgeous design. Dine at its brasserie, where I often have business lunches for quality and price. **les-haras.fr**

Villa René Lalique, Wingen-sur-Moder
Worth the 45-minute drive northwest from Strasbourg: relax in sumptuous rooms fitted with Lalique crystal decor. Built by René Lalique in 1920, the house has been refurbished as a boutique hotel. Dine at the superb two-star Michelin restaurant, and don't miss the Lalique museum! **villarenelalique.com/l-hotel**

RESTAURANT

La Fourchette des Ducs, Obernai
Originally built by Ettore Bugatti, founder of the eponymous car company, to receive customers in appropriate style, the gorgeous interior mirrors peerless cuisine and wine at this two-star Michelin restaurant, ideal for dinner after having trekked through southern Alsace wine country. **lafourchettedesducs.com**

Hôtel Cour du Corbeau, Strasbourg

The beautiful village of Hunawihr, in southern Alsace

(LONG) WEEKEND OPTIONS FROM STRASBOURG

After marvelling at its beautiful interior, climb the 330 steps up the cathedral of **Notre-Dame de Strasbourg**, the sixth-tallest church in the world.

Near European institution buildings – Strasbourg is headquarters for the Council of Europe and hosts monthly EU Parliament sessions – visit the **Parc de l'Orangerie**, the city's oldest park, with sprawling floral grounds surrounding the Pavillon Joséphine, a fine example of early 19th-century neo-classical architecture. Visit the mini zoo, picnic near the lake and waterfall or just take a relaxing walk – or even invigorating jog.

Wine bar urge? Strasbourg's **Ill Vino** on a barge *(illvino.com)* offers a modern, fun ambience – the name refers to the Rhine tributary. Take a deep dive into Alsace wine at **L'Alsace à Boire** (see p85).

Rent a car and visit wineries along the 170km-long picturesque **Alsace Wine Route** (wineroute.alsace), but make sure you book any visits in advance.

Shop at **Beauvillé** *(beauville.com)*, famous producer of luxurious tablecloths and household linens, located in Ribeauvillé, where you'll also find celebrated wine producer Domaine Trimbach.

Le Cerf, Marlenheim
French newspaper Le Monde calls the traditional choucroute of this one-star Michelin restaurant the most successful in France, to which I can attest, as well as to its excellent wine list. A 20-minute drive west from Strasbourg, it is also convenient for its four-star hotel, with a small spa including sauna and hammam. **lecerf.com**

Les Plaisirs Gourmands, Schiltigheim
Pleasingly economically priced one-star Michelin in this Strasbourg suburb that's known for its jazz concerts with international stars. €380 for a delicious six-course lunch for three, including two bottles of wine and tips. **les-plaisirs-gourmands.com**

Mademoiselle 10, Strasbourg
Casual ambience in the city centre, with a view of the Ill tributary, this is often a top choice for diplomatic lunches. Try the €45 lunch menu. **mlle10.fr**

SHOPS & MARKETS

Les Boutiques de l'Aubette features shops at Strasbourg's Place Kléber in a splendid 18th-century building. Although bombed in the Franco-Prussian war and reconstructed several times since, a more definitive 2008 renovation has made it a Strasbourg shopping highlight, in contrast to the rather ugly Les Halles nearby... **laubette.com**

La Petite France
Sometimes called mini-Venice, because the Ill tributary predominates, Strasbourg's Petite France quarter is invaded by tourists. But it is well worth visiting for its medieval architecture, charming narrow paths, and restaurants, shops and boutiques, from Le Goût du Terroir cheese shop to the lovely Galerie Pascale Froessel. Snack on tarte flambée, washed down with Riesling – or Fischer beer. **visit.alsace**

Strasbourg Christmas market
The month-long festive event (Le Christkindelsmärik) starts in late November and features a magnificent Christmas tree and ubiquitous holiday decorations that never fail to enchant. Between sips of spicy hot mulled wine, take care to avoid tourist traps among the temporary wooden holiday kiosks that dot the city. If you like foie gras, buy the amazing Champagne and pepper foie gras at the stand in front of the gorgeous Maison Kammerzell in Place de la Cathédrale.

ALSACE WINE: THE GRAPES

Riesling is king, but styles – and terroirs – vary. Either way, dry Riesling from quality producers matches any great dry white in the world.

Gewürztraminer is slightly on the sweeter side, with low acidity, at its best conveying exuberant lychee, rose and ginger aromas. Pairs well with strong cheese or spicy foods.

Pinot Gris tends to be, as the French say, entre deux chaises, as even 'dry' it often has sweetish overtones and usually a thicker texture than Riesling.

Sylvaner can be superb from older vines that lend depth to what more usually is light dry white.

Pinot Blanc tends to be soft and rather delicate – and is often underrated.

Muscat, when vinified dry, pairs well with white asparagus, or enjoyed on its own as an aperitif.

Mélanie Pfister

For reds, Lucas Alessandri of the Strasbourg Léonor hotel restaurant praises **Pinot Noir**. I have been sceptical, but climate change seems to be 'helping'. Never have I enjoyed such fine Alsace Pinot Noir as in recent years, but I would avoid the house reds at most locations. As Burgundy prices scrape ever higher skies, try Pinot Noir from estates such as Domaine Albert Mann and Mélanie Pfister (see p13), for example. I had a very good Domaine Boeckel, Oberpfoeller 2019 at Léonor that cost only €46 a bottle, restaurant price.

(+33 [0]3 89 27 11 32), you cannot help but admire the adjacent – and steep (45°) – granite Sommerberg vineyard. Oriented directly south, and rising nearly 400m in altitude, it is the source of splendid Rieslings that you can taste at the estate. It is hard to beat the beautiful aromatic minerality and subtle citrus notes encountered in these wines. Try the old-vine Sylvaner and the excellent crémant, too.

No visit to vintners in Alsace would be complete without seeing Famille Hugel in Riquewihr (*m.hugel.com*), a beautiful medieval town along the Alsace Wine Route. Founded in 1639, the estate is still 100% family owned and is now managed by its 12th consecutive generation. A wide range of wines, including Alsace Grand Cru, are sold worldwide: widely recognised for their high quality. Open every day of the year, the cosy tasting room awaits you at the foot of cobbled streets near wood-timbered shops in the village centre. Surrounding vineyard slopes accentuate the charm of the setting. Inside, don't miss a tour of the cellars, full of large oak barrels dating back to the 18th century. Indeed, the 'S. Caterine' cask from 1715 has been named the world's oldest working cask by the Guinness World Records, according to the

'A 20-minute drive from Strasbourg and you're in wine country'

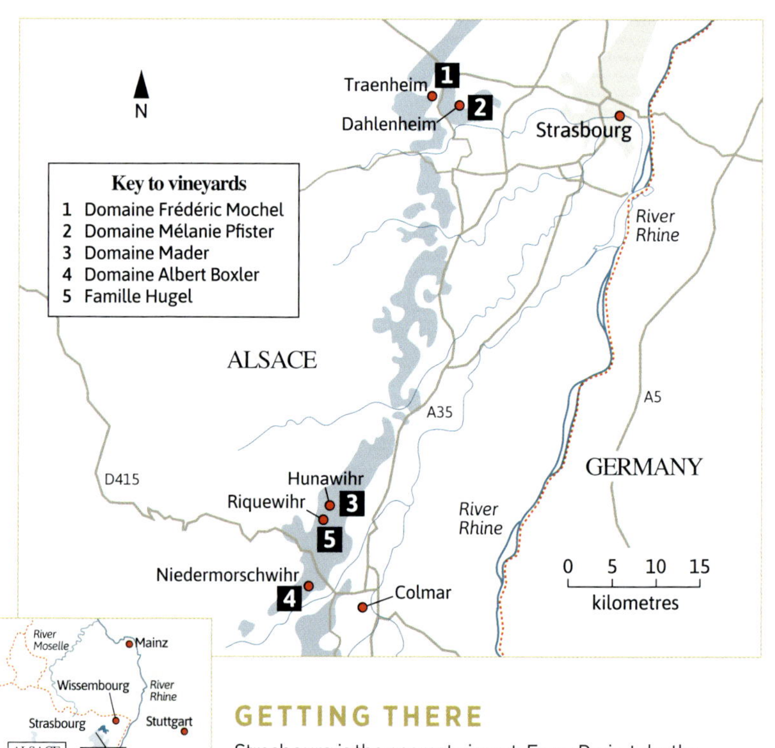

GETTING THERE

Strasbourg is the nearest airport. From Paris, take the high-speed TGV train running daily to Strasbourg, with an average journey time of around two hours.

Riquewihr town as seen from the Famille Hugel Schoenenbourg vineyard

estate. And if you missed any wineries in your tour, back in Strasbourg L'Alsace à Boire is dedicated to Alsace wine (*alsaceaboire.fr*), although I am surprised that it took so long for the region's capital to have such a wine bar (it opened in 2021). But now locals and tourists can discover most Alsace Grand Cru appellations without having to leave Strasbourg. There are some excellent sparkling wines, too – industrial versions of Alsace crémant can be downright average, but many producers excel. Just ask the bartender.

FACT FILE: ALSACE

PLANTED AREA 15,600ha
APPELLATION AP officially instituted in 1962; AP Alsace accounts for about 70% of production
ALSACE GRAND CRU AP Introduced in 1975, progressively expanding to 51 sites classified as such, accounting for about 5% of total production
AUTHORISED GRAND CRU GRAPES Gewürztraminer, Muscat, Pinot Gris and Riesling, but blends are allowed in Altenberg de Bergheim and Kaefferkopf, and Sylvaner is permitted only in Zotzenberg
CRÉMANT D'ALSACE Sparkling made by the traditional method, accounts for about 25% of production
VENDANGES TARDIVES On wine bottles designates sweet wines made from grapes picked when overripe
SUB-REGIONS TWO The southern Haut-Rhin, which counts 37 grands crus, and the northern Bas-Rhin with 14

BORDEAUX CHATEAUX

The estates of Bordeaux have it all: history, art, stunning landscapes and, of course, great wine. With a bit of planning, you can experience the best the region has to offer.

Touring the estate at Château Carbonnieux

CHATEAU
CARBONNIEUX

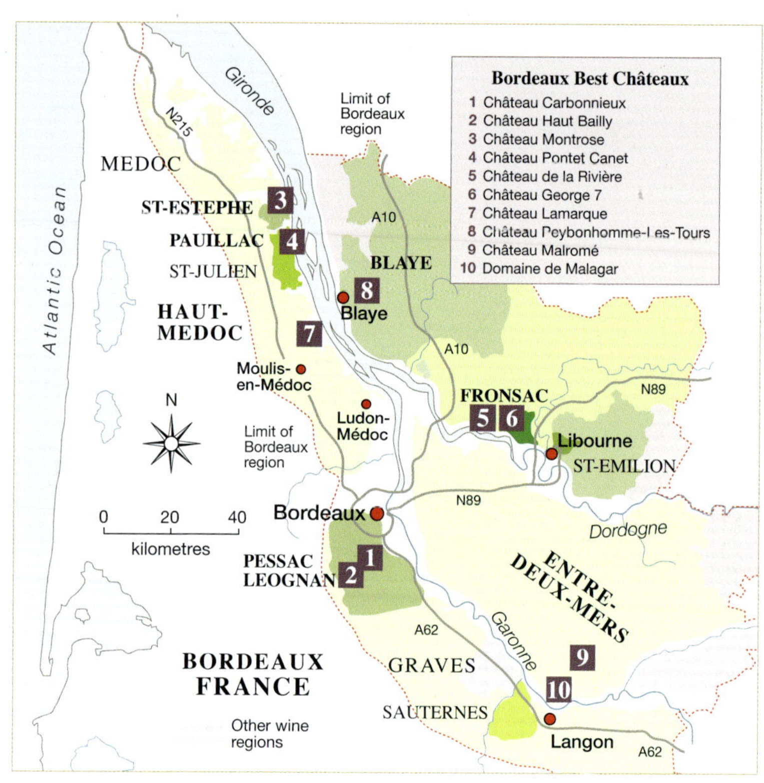

Bordeaux remains a favourite destination for wine-touring holidays. But with 6,500 estates here, where do you start? Selecting châteaux that are near to each other but show differing approaches can be a fun and rewarding way to plan your trip, as with every visit you put a new piece of the puzzle into place. Each pair of châteaux described here can be visited in one day, allowing for a long, slow lunch in between, or a walk through the nearby countryside.

White vs red

CARBONNIEUX & HAUT-BAILLY

Although Bordeaux is 90% red wine, there are many brilliant white wines to discover in the region, and splitting your day between both is a fascinating reminder that for much of the 20th century, Bordeaux made more white than red. You will find examples of white wines all over Bordeaux, and you could easily do this pairing in several appellations (try Château Thieuley/Château de Reignac in Entre-Deux- Mers, or Château Chantegrive/Château de Portets in Graves) – but for Bordeaux's best- known whites, head to Pessac-Léognan.

'Although Bordeaux is 90% red wine, there are many brilliant white wines here as well'

MORNING

Château Carbonnieux

PESSAC-LÉOGNAN CCG

Owned by brothers Eric and Philibert Perrin (no relation to the Château de Beaucastel Perrins in the Rhône), this wonderful estate is pretty much split down the middle into the production of red and white wine, with 50ha of red and 42ha of white – more of the latter than any other Pessac-Léognan estate. The resulting white is a brilliant wine full of creamy concentration, made with the two best-known local varieties, Sauvignon Blanc and Semillon. It's an excellent estate to visit: the property dates back to the 13th century, and for a long time was owned by Benedictine monks known for their exceptional white wines. In 1786, Thomas Jefferson visited and left his mark by planting an American pecan tree, which still stands today.

There are a range of tours available here, from the 'Classic' €10 tour with a tasting of two wines, to the 'Prestigious' tour for €20, with a tasting of three wines and a food platter. A food and wine matching workshop is also on offer, pairing five cheeses and three wines for €22. Open Monday to Friday all year, plus Saturdays from May to October.

carbonnieux.com

AFTERNOON

Château Haut-Bailly

PESSAC-LÉOGNAN CCG

Only red wine is made at this 30ha American- owned estate, and it is easily some of the best in the whole of Bordeaux. It's just along the road from Carbonnieux, so making a visit between the two properties is easy and enjoyable on foot or by bike if you are staying in the area. The wine is classically graceful, full of softly curling woodsmoke, tobacco and rich black fruits – the tasting here is a must. The Haut-Bailly estate's history can be traced back to at least 1461, and the current

Château Haut-Bailly

château is from the 19th century, standing in contrast to the sleek modern cellars. It boasts a well- stocked boutique that sells books, picnic gear and a ton of interesting gifts. For special occasions, you can arrange private dining with the on-site chef. A new winery is currently under construction.

A number of different visits are on offer, ranging from €20 for one hour, to €50 for the 90-minute 'Collector's' session. All include a tour and tasting. ***haut-bailly.com***

Cutting edge vs low tech

MONTROSE & PONTET-CANET

These are two of Bordeaux's most celebrated and iconic names, both producing incredible wines but reaching their goals by entirely different routes. A day spent visiting one and then the other is eye-opening.

MORNING

Château Montrose

ST ESTÈPHE 2CC

Head up to St-Estèphe, where you'll find Château Montrose along the banks of the Garonne river, the far end of its vineyards practically grazing the fishermen's huts along the river. One of the most impressive châteaux to visit in Bordeaux, this 90ha estate is at the cutting edge of vineyard technology with a stunning 10,000m² cellar and luxurious design touches at every turn. It is also one of the greenest châteaux in the region, with geothermal technology, a permaculture orchard, a system for capturing the carbon dioxide produced during fermentation and the use of electric tractors in the vineyard.

In the vineyard things are obsessively tracked and recorded to ensure precision viticulture. The number of plots, for example, stood at 24 when director Hervé Berland arrived in 2011. Today there are 110, the result of dividing and sub-dividing to ensure that all the tiny differences between the plots are respected throughout the growing season and at harvest. You will leave astonished by how much expense, effort and expertise goes into making the greatest wines of Bordeaux.

There is no fee to visit, but you do need to book your tour in advance. ***chateau-montrose.com***

The cellar of Château Montrose

AFTERNOON

Château Pontet-Canet

PAUILLAC 5CC

Just a 15-minute drive south and you cross over into Pauillac, home of the iconic classified estate of Pontet-Canet. You'd be forgiven for thinking that you've stepped back in time, as owner Alfred Tesseron and director Jean- Michel Comme have gone resolutely old- school. Farming is organic, as at Montrose, and also biodynamic, with all biodynamic preparations made on-site. There are even eight Breton draft horses used for vineyard work across at least half of the 81ha estate – if they aren't out working, you can visit them in their well-kept stables. Cellar work is equally traditional. Everything is manual, from destemming and sorting at harvest, to filling cement vats made out of the sand, clay and gravel taken from the surrounding land during the construction of a cellar extension. No electricity is allowed anywhere near the vats, except for LED lighting, and everything is powered by geothermal energy.

There is no fee, but visits here are typically reserved for professionals only, though they may be extended to wine-tasting groups, sommeliers and collectors – it's definitely worth trying. ***pontet-canet.com***

A Breton draft horse in the vineyard of Château Pontet-Canet

Château de La Rivière stands on 100ha of parklands

Historic vs newly created

DE LA RIVIERE & GEORGE 7

Over its 2,000 years of winemaking history, Bordeaux has continually reinvented itself, helped by constant innovation and new arrivals. Nothing brings that home more clearly than a day visiting one of its oldest and one of its newest estates. And Fronsac on the Right Bank is a lovely place to do this, with views over the Dordogne and Isle rivers.

MORNING

Château de La Rivière

FRONSAC

Among the oldest châteaux in Bordeaux, this stunning property was constructed in 1577 by Gaston de l'Isle on the remains of a defensive camp built by Charlemagne. It's hard to miss, as it stands tall over the countryside on 100ha of

Château George 7 is a recent addition to Bordeaux

Sally Evans, owner of George 7, cycling in Fronsac

parkland and gardens that are worth a visit. Best of all are the 8ha of limestone caves that are still used for ageing the wine. They offer a brilliant way to get up close to the limestone terroir that dominates not only Fronsac but also St-Emilion, Castillon and beyond. A range of visits are geared to different audiences, including families. You can even order a picnic to eat in the beautiful courtyard. Although it's a French-run estate, the owners are Chinese, so you might also want to try the pu'erh tea ceremony (€25) to learn the history of this traditional tea, with a tasting.

Open Monday to Friday all year, plus Saturdays from May to October. Visits are by appointment only. A cellar tour and tasting costs €10, while a cellar tour and a tasting of individual grape varieties to show Right Bank and Left Bank differences is €25. ***chateau-de-la-riviere.com***

AFTERNOON

Château George 7

FRONSAC

For something completely different, head to the tiny Château George 7, created only a few years ago by British owner Sally Evans, who changed career to become a winemaker armed only with a WSET Diploma and a positive attitude. It's not easy to find completely new estates in Bordeaux, and it's fascinating to hear about her journey: finding a run-down farm, converting it into a house and equipping an empty cellar building. As for the château name, George means 'tiller of soil' in Greek; it's also a reference to the English patron saint. And the 7? It's possible that Prince William's son George may one day take the name George VII, so Evans used this part of the name to 'pay tribute to the old while venturing into the new', as she puts it. The château opened for business in 2017, although the first vintage was

Old oak vats in use at Château de Lamarque

almost entirely wiped out by frost, making 2018 the true inaugural year. You get to see real hands-on winemaking here – with a bit of help from her consultants, Evans picks the grapes, prunes the vines, and lugs the barrels around. Thank goodness it's only 3ha at this stage. She is also very happy to share her experiences and motivations with others who might be looking to get into winemaking. She can also help you arrange an electric bike tour around Fronsac – particularly useful, as there are some steep slopes in this appellation.

Tours start from €8 (and are free for under-18s) and are held most days from April until October. ***chateaugeorge7.com***

Left Bank vs Right Bank

LAMARQUE & PEYBONHOMME-LES-TOURS

It's not easy to take in both banks in a single day without quite a bit of driving, but this would be a novel way to do it – heading from AP Haut-Médoc over to AP Blaye via the ferry that runs between Lamarque and Blaye. You can take your car on board the ferry. Refer to the tourism website Bernezac for timetables *(bernezac.com)*.

MORNING

Château de Lamarque

HAUT-MÉDOC

One of the most amazing châteaux in the Médoc, located between Margaux and Pauillac, Lamarque should be far better known to wine tourists. It's definitely worth spending a full morning here as there is so much history to soak in, and extremely charming owners to take you through it. The château itself is, in part, 1,000 years old, built originally as a fortress by Garsion de Lamarque to counter the lingering threat of Viking raids. It is perfectly preserved, with a 13th-century keep where the family's private wine collection is stored. A former chapel dates back to the 11th century, and even the winery contains a row of well-preserved old oak vats, which conceal a more modern interior. The current owner, Pierre-Gilles Gromand-Brunet d'Evry, is a direct descendant of Garsion de Lamarque. It's an excellent Haut-Médoc wine to boot. Visits and tastings are by appointment only. ***chateaudelamarque.fr***

AFTERNOON

Château Peybonhomme-Les-Tours

BLAYE

You're going to want to head to the Citadelle de Blaye when you disembark from the ferry, as this UNESCO World Heritage fortress built by renowned military engineer Vauban is one of the best preserved in France. There is a small vineyard there, and you can attend a wine tasting in the Cellier des Vignerons. However, Blaye is also a great place to visit an estate, as lots of owners live on site. I would recommend Peybonhomme-Les-Tours in the commune of Cars, overlooking the estuary. It is certified biodynamic, with great wines produced by the

Château Peybonhomme-Les-Tours overlooks the Gironde estuary

welcoming Bossuet-Hubert family. There are several wines to look out for, particularly amphora-aged bottling Energies and a sparkling blanc de noir – all low sulphur and showcasing the many interesting developments happening in Bordeaux with a more hands-off, natural approach to winemaking. Visits are by appointment only. ***hubert-vigneron.com***

Literary vs artistic

MALROME & MALAGAR

If you are a fan of books or art, there are two estates that give you the pick of both, about 50km to the southeast of Bordeaux on the edge of Entre-Deux-Mers.

MORNING

Château Malromé

BORDEAUX SUPÉRIEUR

Art lovers shouldn't miss this 45ha estate at St-André-du-Bois – it dates back to the 14th century and has been delivering harvests without stopping through wars and revolutions ever since. Best known for its connection to the artist Henri de Toulouse- Lautrec, it was owned by his mother and he died here in 1901. It's worth a visit for its Toulouse-Lautrec exhibition – you can walk through his apartments and see some of his original sketches, including 19th-century graffiti – but also for its contemporary art foundation. The new Franco-Vietnamese owners, Kim Valéry Huynh and his daughters Mélanie and Amélie, hold regular exhibitions. There is also a restaurant, Adèle (named after Toulouse-Lautrec's mother) – an outpost of the popular Claude Darroze restaurant in Langon – that is open Wednesday to Sunday and has an excellent Sunday brunch.

Domaine de Malagar was the home of renowned author François Mauriac

The art gallery at Château Malromé

A guided visit is €12, including a tasting. Access to the Toulouse-Lautrec apartment is available only through the guided tour. ***malrome.com***

AFTERNOON

Domaine de Malagar

CÔTES DE BORDEAUX

Although you can see vines around this property near St-Maixant, the wine sold here is produced by Jean Merlaut in a winery at the bottom of the hill, close by but not open to the public. Called Château Malagar, it is made in Côtes de Bordeaux red, AP Bordeaux white and rosé, and sweet white Premières Côtes de Bordeaux. In addition to buying wine at Malagar, you can also check out an exhibition about the life of one of Bordeaux's most famous authors, François Mauriac, winner of the Noble Prize in Literature in 1952, who used to live here. But one of the best things that you can do when visiting Malromé and Malagar is walk – there is a footpath that connects the two estates. The 7km journey is open to any enthusiastic hiker, and twice a year the estates host a joint celebration called Sur le Coteau des Artistes, where you start at Malromé with a visit and a tasting, then walk over to Malagar for a picnic followed by a visit, then return by a different footpath. It's a full day, but a wonderful way to get to know the beautiful landscape around the Garonne valley in this part of Bordeaux.

Open daily from February to November. A guided visit is €8, including a tour of the house, the Mauriac exhibition and the park. ***malagar.fr***

COGNAC

Strike out to the western French department of Charente, a region carved by a winding river, dotted with châteaux and rich with opportunities to experience the most illustrious of French spirits.

The Hennessy shuttle boat on the river Charente

Sweeping from above the Atlantic ocean-kissed shores of La Rochelle, down the Gironde estuary towards the fêted vineyards of Bordeaux – and reaching far inland through forest, rivers and wheat fields – the Cognac region at first seems mind-bogglingly vast. But while this wide expanse may be one of France's largest grape-planted areas, the good news for travellers is that there's a clear first-timer's itinerary to discovering its liquid treasures. Simply make for the heart of the action in the premium central sub-regions of Grande Champagne and Petite Champagne, and the twin spirit-soaked communes of Cognac and Jarnac, for a long weekend of tastings, cellar tours and laid-back exploring.

Covering some 13,000ha in the centre of the Cognac region – and bordering Cognac town – Grande Champagne has highly chalky soils relative to other sub-regions and picturesque vistas of low vine-lined hills. With its white stone buildings along the squiggling river Charente, Cognac town makes a fine base for your explorations, both for its proximity to major houses and its smattering of lovely sights, from half-timbered homes to old cloisters. Conversely, sleepier Jarnac, a 10-minute train ride or 20-minute drive away on the borders of Petite Champagne, has sun-flecked spaces dominated by the grand frontage of Courvoisier *(courvoisier.com)* and unassuming corners housing the likes of Hine (hinecognac.com) and Louis Royer *(louis-royer.com)*. Beyond these two communes, countless other small villages radiate outwards with cellars and tasting rooms both big and small, ripe for discovery if you have the time.

Tasting experience

Given the punchier nature of a Cognac tasting in comparison to wine, you'll likely only want two or three stops a day – which means being selective among the 280 or so regional producers. Wherever your stylistic preferences lie, it's almost mandatory to begin with the house that's synonymous with the Cognac category: Hennessy (hennessy.com; see 'My perfect day', opposite). Responsible for about 40% of all regional production, the LVMH-owned brand's rambling facilities are set over a series of vast buildings in the heart of Cognac. Tours come in every shape and size: join a shuttle boat across the river Charente to discover Hennessy through a digital art installation, sampling the characteristically spicy VSOP on a two-hour group tour (€29 per person, times vary so check and book ahead); or cycle through the vineyards ahead of a superb picnic lunch from the kitchens of chef Thierry Verrat, owner of one-star Michelin restaurant La Ribaudière (€130 per person, €65 picnic supplement).

A 15-minute walk away from Hennessy, and somewhat more modest in scale, family-owned Bache Gabrielsen (*bache-gabrielsen.com*) has its headquarters in a small townhouse on tranquil Rue Louis Dominique. In the historic offices where respected Master Blender Jean-Philippe Bergier works his magic, old wooden cupboards line walls and samples from local bouilleurs de cru ('grower-distillers') clutter desks awaiting consideration for future assemblages ('blends'). Join a tour of the onsite cellar, spotting precious demijohns of old vintages and sampling through the smooth and complex range (free of charge). If you've got the cash, you can even buy your own barrel for bottling when the time is right.

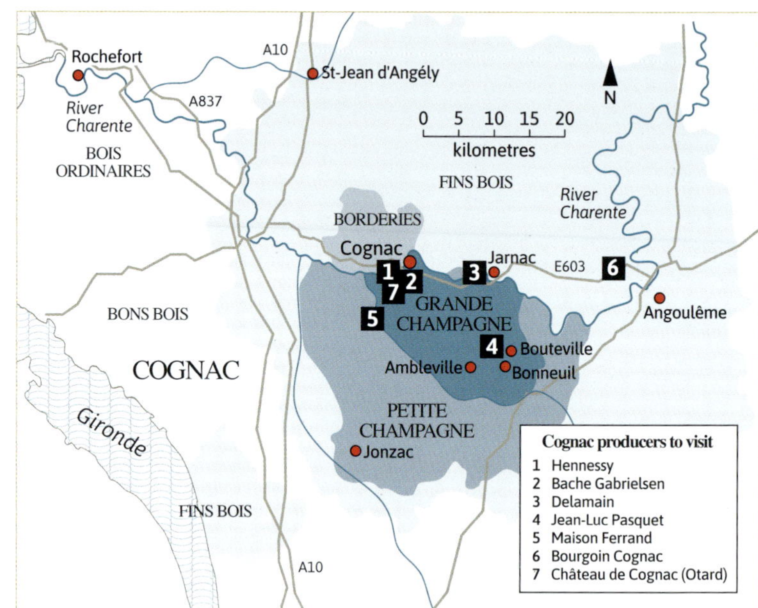

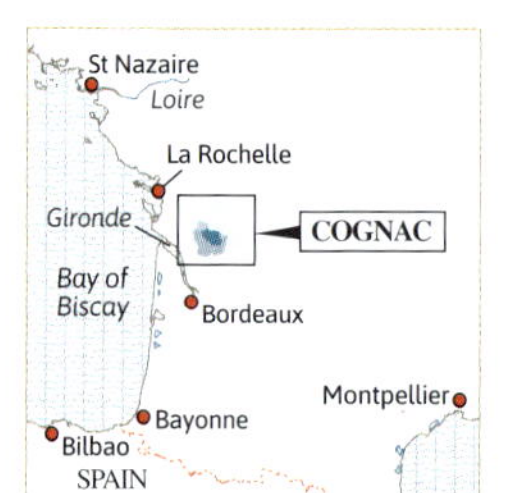

GETTING THERE

The nearest well-served airport and major city is Bordeaux-Mérignac (130km south), but flights also go into La Rochelle (110km northwest, on the coast) and Limoges (140km due east inland). Or take a train to Cognac or Jarnac station from Paris in three to four hours.

Over in nearby Jarnac (see map, above), your essential stop is the highly respected Delamain (*delamain-cognac.com*), blending elegant, long-aged Cognacs from an elite selection of farmer-growers around the region with spirit distilled from its own vineyards. The time-warp ageing rooms, filled with barrels and demijohns, are among the most atmospheric to be found in all of Cognac, and private cellar dining experiences with tastings of signature Pale & Dry XO, as well as highly limited Pléiade collection bottlings, let you soak it up to the fullest.

Frédéric Bourgoin, whose family vineyards provide a friendly stop-off

And beyond

Now, venture into the countryside. Set just outside scenic castle town Bouteville among spectacular, meandering country roads, Jean-Luc Pasquet (*cognac-pasquet.com*) is a small husband and wife set-up

'Time-warp ageing rooms, filled with barrels and demijohns'

The atmospheric ageing cellars of Delamain in Jarnac

MY PERFECT DAY IN COGNAC

MORNING

After a quick spin around the morning produce market in **Cognac** town, kick off your introduction to the long-aged Cognacs at Hennessy, by far the region's largest producer. Its 90-minute Initiation visit (€25 per person) is ideal for beginners who want to learn more about the spirit's production process, but connoisseurs will be more enthralled by the four-hour Grape to Glass masterclass (€500). Afterwards, depending on time, visit Bache Gabrielsen for a brief tour and tasting (p96) or try the **Château de Cognac** *(chateaudecognac.com)*, birthplace of François I and now a tasting room for Cognac Baron Otard.

LUNCH

Nibble on market-fresh fare at Poulpette *(see 'Address book', p31)*, a few minutes' walk away across the Charente river. Concise seasonal menus of elevated bistro dishes are paired with wines from small minimal-intervention winemakers such as Languedoc's Domaine Peyrus and Château Thivin in Beaujolais.

AFTERNOON

Hop on the train for the 10-minute ride to neighbouring **Jarnac** – whizzing past rows of grape vines – to get a view of the region's other must-see centre. You should have booked ahead if you want to explore the atmospheric ageing rooms of Delamain; access is limited. But you will also find a newly renovated tour area in the grand Courvoisier building, set opulently on the riverside. Book the 1.5-hour tour and tasting, then enjoy a pick-me-up coffee in a quiet square in the sun and soak up the laid-back Gallic atmosphere.

EVENING

Take the train back to Cognac for dinner at restaurant Les Foudres, set among 100-year-old barrels in the centre of town at Hôtel Chais Monnet. Afterwards, the hotel's 1838 Jazz Bar

Jean-Philippe Bergier, Bache Gabrielsen cellar master

awaits with its unrivalled selection of Cognacs – there are more than 220 pours to choose from. Ask the knowledgeable bartenders to put together a tasting flight for you to sip while you enjoy some live music.

operating as both bouilleur de cru and blender. Working organically across about 15ha, mostly of Folle Blanche and Ugni Blanc, they encourage biodiversity on site and distil in their historic farmyard building. Settle into the tasting room sofa with American-born Amy Pasquet and savour the appley four-year-old L'Organic or spicy, leathery 10-year-old.

An innovator of another breed in the commune of Ars, Maison Ferrand (maisonferrand.com) was founded in its current form by Burgundian Alexandre Gabriel – the brain behind the refined and dynamic Plantation Rum range – in 1989. The focus is not only on quality but on crafting unique, consumer-friendly bottlings such as 10 Générations, aged partly in ex-Sauternes casks, and Double Cask Réserve, rested in barrels that formerly held Banyuls. There is not officially a visitor centre, but some visitors are welcomed by appointment; enquire via the website.

'You'll discover a range of delicious, and often excellent-value, aged eaux-de-vie that show exactly why the spirit is held in such high esteem'

For one more boutique stop-off, consider Bourgoin Cognac (bourgoincognac.com), just outside Angoulême, a Charente city you may recognise for its starring role in Wes Anderson's film The French Dispatch. The 'micro barrel' XO range produced from family vineyards is decadent, nutty and accomplished – and a first-rate bottled verjus (unripe grape juice) for cooking or cocktail making is also produced. Join a 90-minute 'safari' in the vines (€75 per person) for the most comprehensive experience.

Ultimately, wherever you choose to stop off in Cognac, you'll discover a range of delicious – and often excellent-value – aged eaux-de-vie that show exactly why the spirit is held in such high esteem. And unlike many wine regions, Cognac can be an enticing prospect in winter – not only for the pleasure of sipping a fragrant VSOP by a roaring hotel fire, but because distillation takes place between October and March (Jean-Luc Pasquet, for example, is even known to invite some visitors to take part in the process if booked as a special package: contact directly to enquire).

Saying that, if you want sunshine, spring is the loveliest time. The leaves are green on the vines, the sun bathes walks along the Charente river and the mood is jovial. Even if you've never been much of a Cognac drinker, a few days in this fascinating region will quickly have you entranced.

Cycling tours are a popular way to explore the Cognac vineyards

The luxurious towered Domaine des Etangs

YOUR COGNAC ADDRESS BOOK

ACCOMMODATION

Domaine des Etangs

East of Cognac town, this luxurious, turreted château hotel is nestled on rambling farmland with lakes, cattle and large-scale sculpture. Decor is contemporary and art-forward, and an honesty bar stocks sublime, vanilla-noted Bourgoin Cognac. **aubergeresorts.com/domainedesetangs**

Hôtel Chais Monnet & Spa

Locations don't get more prime: right in the heart of Cognac town centre. Set within former cellars reworked into a sculptural, contemporary vision of glass and steel, also ticking boxes for its Michelin-starred restaurant *(see 'Les Foudres', below)*, elegant spa and list of local experiences, ranging from private tours to bike rides through vineyards. **chaismonnethotel.com**

Le Relais de Saint-Preuil

For a cosy traditional manor house atmosphere, coddled in vines in between Cognac and Angoulême. A suntrap pool awaits for post-tasting relaxation in summer, while roaring open fires warm in cooler months. **relais-de-saint-preuil.com**

RESTAURANTS

Le Verre y Table

A few steps from Jarnac rail station, this modernist, conceptual dining space dishes up the likes of Charron mussels with pineau, or trout with squid ink and lemongrass. A €29 set lunch menu is great value, and any time of day the extensive Cognac menu is welcome. **leverreytable.fr**

Les Foudres

Located in a former ageing warehouse, this one-star Michelin restaurant has distinct local flair. Menus include the likes of caviar butter with seaweed brioche, asparagus with coffee and Martell Cognac creme, or stone bass with cream bottarga and potatoes. **chaismonnethotel.com/savourer/les-foudres**

Poulpette

With a concise menu and contemporary platings, this Cognac town restaurant, located a short walk from the Charente river, is a favourite with locals thanks to its regularly changing market menu. Wash it all down with a wine list featuring cult producers. **poulpette.squarespace.com**

Menus at Poulpette change weekly

THINGS TO DO

Abbaye de Bassac

With more than a thousand years of history, this rambling stone abbey outside Jarnac gives insight into Benedictine lifestyle and French architectural styles from Romanesque to Baroque. Its wealth of delightful outside spaces are perfect for wandering around in the spring and summer months. **abbaye-de-bassac.com**

Le Baume de Bouteville

This artisan vinegar producer uses locally grown grapes in its production and ages its balsamic-style concoctions in ex-Cognac barrels. Join a tour and tasting (with fresh oysters, if desired) at its boutique headquarters under the shadow of Bouteville's château. **bouteville.com**

River cruises

Tour the snaking river Charente aboard electro-solar boat Bernard Palissy III, setting off west of Cognac town and taking in cobbled settlements, farmers' fields and sun-sparkled waters. **croisieres-palissy.fr**

CONDRIEU

Too richly flavoured for some, nectar to others, Viognier finds its apotheosis in Condrieu. Here in this far-northern Rhône region, up-and-coming winemakers are meeting with aplomb the challenges posed by shifting tastes and rapidly changing climate.

Stéphane Montez of Domaine du Monteillet looks down on the bridge over the Rhône at Chavanay

'When Viognier is grown on this particular sequence of airy granite slopes, it takes on a salinity and intensity'

In the 1960s, Viognier nearly died out completely. Many Decanter readers would be horrified by this scenario, but certainly not all – it is, after all, one of the most divisive of grapes. Its birthplace is Condrieu, and the new president of the appellation, Pierre-Jean Villa, is under no illusions. 'People either like it or they don't.' he says. 'It's not universally enjoyed like Chardonnay.' But those that like it, love it. To him, 'Condrieu is magic – but fragile'. In the face of climate change, he's helping a new generation of winemakers adapt their winemaking to create a fresher style of wine.

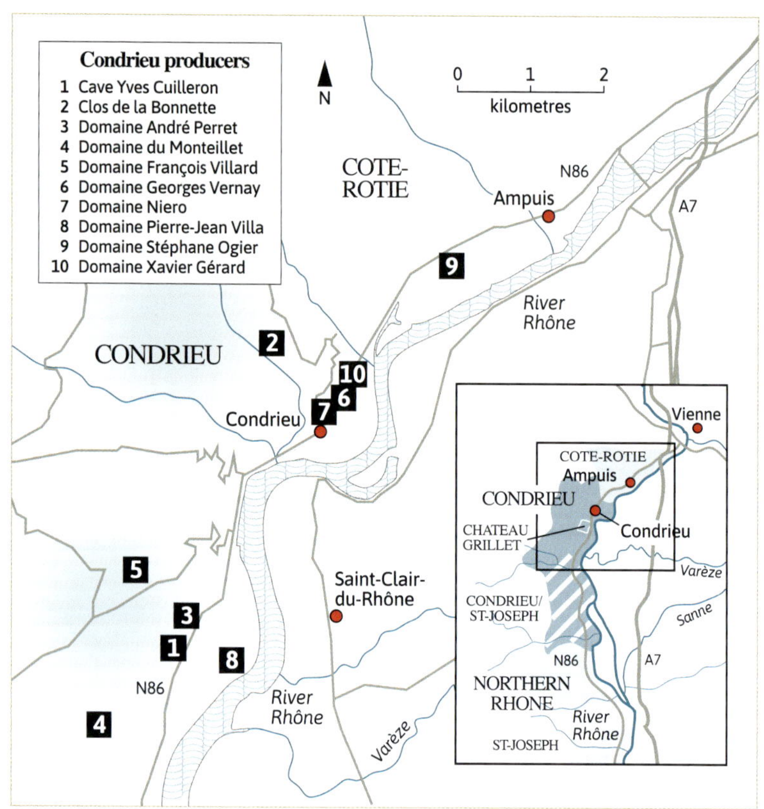

Founding fathers

Today there are more than 16,000ha of Viognier grown around the globe, but it wasn't always so widespread. It originates in the northern Rhône, where it has been cultivated for more than 2,000 years. But the late 19th and early 20th centuries were

Domaine François Villard

FACT FILE: CONDRIEU

AP CREATED 1940; one of the eight northern Rhône crus
WINES 100% still white, mostly dry but a small amount of sweet Condrieu is produced
GRAPES 100% Viognier
CLIMATE Continental: hot summers, cold winters, strong north wind
PRIVATE ESTATE 87
ANNUAL PRODUCTION (2022) 215ha producing 7,961hl (average yield 37hl/ha)
AMOUNT EXPORTED 32%
VINEYARD SURFACE CERTIFIED ORGANIC 14%
VINEYARD SURFACE CERTIFIED HVE 31%
TOP RECENT VINTAGES 2021, 2020, 2016, 2014

[SOURCE: INTER RHONE, 2022]

Christine Vernay

'We need to find freshness – without forgetting we're from the Rhône'

Aurélien Chirat, **Vignobles Chirat**

tempestuous, and after the region was battered by phylloxera and two world wars, by 1965 plantings had dwindled to just 8ha. The late Georges Vernay never stopped believing in Condrieu, however, and he inspired a group of fellow winemakers to replant the region's vertiginous terraces. Together they saved Viognier from extinction and since then it has spread around the world.

Strong character

For anyone studying for wine exams, being served a Viognier in a blind tasting is always a bonus as it's so easily identifiable. But strong characters can be polarising. Some love the variety for its opulent fullness, satin texture and heady fragrance of peach, violet and jasmine. For others, its acidity is too low, alcohol too high, and the perfume and flamboyance are overpowering.

Contrarily, it's possible to like Condrieu even if you're not a huge fan of Viognier. When Viognier is grown on this particular sequence of airy granite slopes – particularly the dark biotite granite at the heart of the appellation – it takes on a salinity and intensity that helps to balance the variety's natural propensity to corpulence.

But the 21st century has produced a new threat – one that can exaggerate Condrieu's personality still further: a rapidly heating climate. It's something of which the new generation of local winemakers are acutely aware.

Aurélien Chirat, who recently took over his family estate after spending time in New Zealand, says 'we need to find freshness – without forgetting we're from the Rhône'.

Leading from the front

When the presidency of the appellation became available, Pierre-Jean Villa hoped that a young wine-grower might rise to the occasion. But he didn't see many hands go up, he says, and he gradually realised that most were too inexperienced for this political role. In the spirit of Georges Vernay, Villa felt a duty to lead from the front, 'until someone from the next generation is ready to take it on'. They're all fortunate: Villa is one of the finest vignerons of his generation.

Villa has identified two main objectives, the first of which is to update and refresh the image of Condrieu – he is currently working with an agency on branding and digital communication.

The second is altogether more crucial: to help local winemakers adapt their viticulture, 'to avoid the wines getting overly heavy and rich in the face of climate change', he says.

Villa is an expert in all things viticultural and has several approaches in mind, such as using massal selection instead of clones (ie, propagating new vines using cuttings from the best existing stock of old vines in a vineyard) and choosing more resistant rootstocks. He's also keen to promote agroforestry (planting trees within and around vineyards), which can help to reduce erosion, increase biodiversity and reduce greenhouse gases – something he's been experimenting with in his own holdings.

Villa points out that Viognier has one major strength in the battle against climate change – it's not averse to hot weather. It's a naturally low-acid variety that finds balance through minerality and positive bitterness. 'We've always made wines at 14% with low acidity,' says Villa.

Fighting the flab

Xavier Gérard took over the family estate from his father François in 2013 and has rapidly gained a strong following for his Condrieu. He says that the new generation of winemakers have been adapting their work in the cellar as well to create a more balanced, food-friendly style.

In the winery, keeping temperatures low during fermentation and very gentle pressing to avoid crushing the skins both help to rein in overly exuberant fruit flavours. Reducing lees stirring during maturation helps to avoid flabbiness. And dialling down on the oak also helps – particularly heavily toasted barrels that add vanilla flavours and oak tannins that 'make the wine even fatter', says Gérard. 'I'm really backing off from small barrels and new oak.'

In order to preserve freshness and acidity in the wines in warmer vintages, Gérard has an ingenious approach to harvesting. He brings in half of his grapes early at just 11.5% potential alcohol with high acidity, then the other half when they're fully ripe: the best of both worlds.

A greener future

When the late Georges Vernay retired after the 1996 vintage, he handed the estate to his daughter Christine, who still leads the estate today. Growing grapes without resorting to herbicide in Condrieu is particularly challenging, since everything on these precarious terraces needs to be done by hand. But Domaine Georges Vernay is now certified organic and they are working towards biodynamics. 'This is a historic estate,' says Christine, 'and we have a duty to lead by example.'

Christine Vernay has proved an inspiration in other ways. Female winemakers were rare 20 years ago, but she has demonstrated that success is possible for the women who are taking over their family estates today. Among them is her daughter Emma. I asked Christine which areas Emma will have to master at the domaine. 'Everything,' she answered.
It's not going to be easy. With Viognier, she'll be growing a demanding grape variety in challenging terrain, and in a chaotic climate. But the new generation in Condrieu benefit from strong leadership, and are adapting their methods to make some thrillingly fresh and drinkable wines.

It's not for everyone, admittedly. But for those of us who love it, nothing compares – just one sniff and we know we're in for a treat.

RIGHT Henri and Isabelle Guiller-Montabonnet, Clos de la Bonnette

10 NAMES TO KNOW IN CONDRIEU

CAVE YVES CUILLERON

For many, Yves Cuilleron is synonymous with Condrieu, but when he originally took over the family estate, in 1987, it was just 3.5ha. Now he owns plots in almost all northern Rhône appellations, totalling 75ha. Winemaking, however, remains notably hands-off, and the results reliably excellent.

CLOS DE LA BONNETTE

Established by Isabelle and Henri Guiller-Montabonnet (pictured, p36) in 2009. Their picturesque vineyards are tucked away in a verdant valley in the hills behind Condrieu town and have always been organic. Their son Antoine is now taking up the reins. Try their vibrant Condrieu Légende Bonnetta (2020, £55.95 Lea & Sandeman) to see why this is one of my favourite new estates.

DOMAINE ANDRE PERRET

When André took over from his father in 1982, the estate was mostly fruit trees. What he hands over to his daughter Marie today is one of the very foremost Condrieu estates. His cuvée Chéry is consistently one of the greatest wines of the appellation – a measured and beautiful expression.

DOMAINE DU MONTEILLET

At wine tastings around France, this estate's stand is always mobbed. It's no surprise as owner Stéphane Montez makes complex and expressive St-Josephs and Côte-Rôties – and some thrilling Condrieu. His cuvée La Grillette from a plot bordering Château-Grillet is particularly concentrated and intense.

DOMAINE FRANCOIS VILLARD

François was training to be a chef when he was bitten by the wine bug and bought his first Condrieu vineyard in 1988. He now has 40ha around the northern Rhône. He makes three expressive Condrieus from different terroirs and his new fourth cuvée, Villa Pontciana, is a selection of his best barrels.

DOMAINE GEORGES VERNAY

Christine Vernay, took over from her father Georges in 1996; she makes three Condrieu cuvées and all are consistently excellent. Her Terrasses de l'Empire (2018, £72.95 Wine Republic) is the most approachable and classically styled. Les Chaillées d'Enfer (2020, £98.75-£120 All About Wine, Fintry Wines, Strictly Wine, VINVM) is more concentrated, from two 50-year-old plots of Viognier. The Coteau de Vernon is a wine of great complexity and freshness that can last for decades.

DOMAINE NIERO

One of the rare northern Rhône estates that produces more white wine than red, and is expert at Condrieu. Established in 1985 by Robert Niero, now maged by his son Rémi. In conversion to organic.

DOMAINE PIERRE-JEAN VILLA

Originally from Chavanay in the northern Rhône, Villa trained and worked in Burgundy before returning home to establish his own estate, which is now certified organic. Villa is an expert in all things viticultural and recently took over as the new president of the appellation. Wines of great balance and elegance.

DOMAINE STEPHANE OGIER

Stéphane's father Michel established the estate in 1983 and within 30 years it reached the top tier of northern Rhône producers. Best known for his single-vineyard Côte-Rôties, but his Condrieus are equally impressive. His Combe de Malleval (2020, £45 Laithwaites) is generous but not exaggerated, while his Vieilles Vignes de Jacques Vernay (no relation to Georges) is a more intense, ageworthy example.

DOMAINE XAVIER GERARD

Unfailingly enthusiastic, Xavier took over the family domaine from his father François in 2013, after a stage at Boekenhoutskloof in South Africa – known for its excellent Syrah. He's one of northern Rhône's most exciting new talents, making fresh and drinkable Condrieu that's now among the very best.

Yves Cuilleron

Pierre-Jean Villa

The Iles Sanguinaires on Corsica's southwest coast, with one of the island's many Genoese-era towers

CORSICA

Going off-peak offers the perfect opportunity to enjoy the captivating scenery and culture of this memorable island, almost as much Italian as it is French, and to seek out gastronomic delights you won't find in the summer months.

Pick any vineyard in Corsica and there is a high chance you'll enjoy a spectacular view as well as excellent wines. You might gaze upon the sea from the vineyards in Cap Corse or admire the wave-shaped La Conca d'Oru mountain in the Patrimonio wine region. Wherever you are on the island, the breathtaking scenery varies, but so too does the soil, ranging from slate to clay-limestone to granite. It means Corsica delivers a huge variety of wine, even with the same grapes being used throughout.

Although the summer sees a huge influx of visitors, those in search of Corsica's vinous and gastronomic assets will find the cooler months bring fewer crowds, along with rewards such as vineyard walks and the chance to try its rustic, seasonal cuisine which isn't available in peak summer – the island's much-loved brocciu cheese comes back into production in October, while the chestnut pulenda (similar to polenta, but made with chestnut flour) and figatellu sausage are eaten over the autumn and winter. A week allows for a whistle-stop tour of the 180km-long island, but a trip of 10 to 14 days allows for a more comfortable pace.

A good place to start is Domaine Devichi (*mlledevichi.com*), which is now in the hands of Marie-Françoise Devichi, the sixth generation and the first woman in the family to run the estate. As well as developing her own modern 'Mlle Devichi' branding for the wines and experimenting with biodynamics and ecological methods, she offers visitors the chance to take guided walks around the 42ha site to learn about the clay-limestone terroir and how the grapes – namely Nielluccìu for reds, Vermentino for whites and Muscat à Petits Grains for dessert wine – grow in the company of broom flowers, wild asparagus and wild pear.

At the centre of the vines is her grandfather's winery and house, shaded by mighty 200-year-old eucalyptus trees. Devichi hopes to one day put it back into use, but for now tastings are offered at her family's newer winery, set in the village of Barbaggio. The Patrimonio appellation, to which Domaine Devichi belongs, was the first in Corsica to be awarded its AP, in 1968, and it has a winning combination of terre et mer in that it is close to the sea with a clay-limestone soil. It lies at the foot of Cap Corse, the mountainous peninsula that juts out to the north. The wines here gain minerality from the slate soil and have a much stronger influence from the sea. This is also reflected in its social history: its inhabitants were sailors and explorers rather than farmers and shepherds as with the rest of the island.

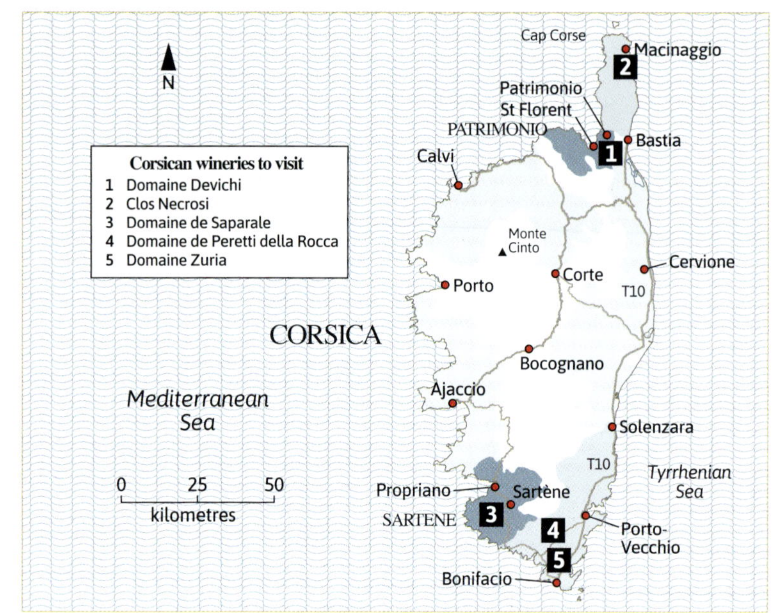

GETTING THERE

Bastia airport, in the north of the island, is less than two hours' drive from the tip of Cap Corse and 30 minutes from the Patrimonio vineyards. Figari airport, in the south, is 30-40 minutes from the vineyards of Sartène and Bonifacio.

Domaine Saparale, near Sartène village

Seafaring legacy

Tucked away in the village of Macinaggio, in the Cap's northeast corner, Sébastien Luigi's family winery Clos Nicrosi (+33 [0]6 11 91 12 15) was established in the 1850s by Luigi's four-times-great-grandfather who had – as many Cap Corsicans did – sailed to the Americas to make his fortune. To show how well he had done, he built their family home, a so-called Maison Américain – these elegant Tuscan-style mansions can be seen throughout Cap Corse and are testimony to a history shared by many other local families. Though Luigi's father and grandfather continued to look abroad to sell, Luigi has scaled back on the exports, preferring to keep the wines more exclusive and sell to the island's best restaurants. One of these smaller-scale wines is rappu, a sublimely moreish sweet red wine unique to Cap Corse made with the Aleatico grape.

In the northwest corner of the Cap you will find Centuri, a fishing village where you can enjoy the local speciality of lobster and langoustine at one of the many quayside restaurants, such as U Cavallu Di Mare (+33 [0]6 18 15 76 31),

MY PERFECT DAY IN SARTENE & BONIFACIO

MORNING

Corsica's villages are ideal for the flâneur; the village of Sartène offers the chance to stroll through narrow medieval streets, up steps and down alleys, before stopping for a café crème in Place Porta. From there it is a 25-minute drive to Domaine Saparale, a beautiful winery overlooking L'Omu di Cagna. It is set in the buildings of a former hamlet where you can see that the estate even had its own police station to protect it from 19th-century bandits. A tour of the winery followed by a tasting of their excellent wines is a joy, while it also gives the chance to discover their new range of natural wines.

AFTERNOON

Just 15 minutes from the winery is **La Bergerie d'Acciola**, a restaurant where the island's charcuterie, meat and cheese producers are celebrated. From there, it's just a 14km detour to one of the island's best beaches, Plage de Roccapina: a strip of fine blond sand lapped by tranquil azure waters. It's a good place to

La Bergerie d'Acciola

Hotel Le Royal

pause before the hour's drive to Domaine Zuria to take a winery tour and a tasting before heading on to Bonifacio.

EVENING

While Bonifacio is best viewed from the sea, you get an incredible view of the limestone coastline and the town's cliff-top houses from the Campu Rumanilu viewpoint a few minutes' drive south. Once you get into the maze of narrow streets, there are galleries and boutiques to explore, while **L'Assaghju** bar (21 rue du Palais) offers the owner's homemade aperitifs made with the island's chestnuts. The restaurant **L'Archivolto** (2 rue Archivolto) is just around the corner serving such dishes as grilled fish and glossy vegetables within its cosy, character-filled interior. Stay at the **Hotel Le Royal** *(hotel-leroyal.com)*, just a few steps away.

with a view of the harbour. Returning south along the twisting roads of the jagged east coast, you come to Bastia. This lively town, where ferries arrive from Italy and France, offers fine food and wine stores, such as Mattei Concept Store and U Paese (*upaese.corsica*), where the island's signature charcuterie hangs from the rafters. Take the newly installed lift from the Quai Albert Gillio up to the town's 14th-century citadel, where you can explore the network of narrow streets, browse its boutiques and pause for dinner overlooking the old harbour.

The beautiful south

The road from Bastia to the Sartène wine region, a 3.5-hour drive to the southwest, takes you right across the island and through the jaw-dropping scenery in its heart. In this region the terroir is granitic and red wines come from the Sciaccarellu grape, as well as Niellucciu and Grenache. A visit to Domaine Saparale (*saparale.com*) near the village of Sartène gives the chance to enjoy a tasting and hear about the domaine's storied past. Saparale was established in 1850 by a lawyer who had explored Africa for years – inspiring the winery's elephant logo – returning home to pursue his dream of developing a self-contained village and vineyards.

'People thought he was mad,' says today's owner Julia Farinelli. 'But he really had a vision to create great Corsican wines.' Sadly, the success was short-lived due to phylloxera and the world wars, but Julie and husband Philippe have restored the estate and now produce natural wines alongside its usual range. A hotel will open in 2024 to add to its luxurious self-catering accommodation, set in the estate's former shepherds' huts.

Indeed, this is how many other wineries in the area, such as Domaine de Peretti della Rocca (deperettidellarocca.com), are using their ancient buildings to welcome guests. As well as his four bergeries, Jean-Baptiste de Peretti runs a chic restaurant overlooking the domaine's vines with a view of the dramatic mountain L'Omu di Cagna.

In the island's far south, the wine scene offers yet another experience. After visiting Bonifacio, a town perched on a high

YOUR CORSICA ADDRESS BOOK

ACCOMMODATION

Domaine Saparale, Sartène
This sophisticated winery in Corsica's southwest has converted three shepherds' huts into luxurious accommodation, each with a swimming pool. Homemade meals and local produce can be delivered. **lehameaudesaparale.com**

Hotel Le Saint Jean, Ersa
Set high above the coast in the village of Ersa at the northern tip of the Cap Corse peninsula, this hotel offers breathtaking views of the coastline. The restaurant offers a good-quality menu and a small selection of local wines. **lesaintjean.net**

La Dimora, Oletta
This chic hotel is set in an 18th-century farmhouse 6km from the small waterfront town of St-Florent, and boasts a beautiful outdoor pool and an excellent restaurant. It is also within easy reach of the Patrimonio vineyards. **ladimora.fr**

RESTAURANTS

A Nepita, Ajaccio
British chef Simon Andrews arrived in Corsica 20 years ago and A Nepita has become one of the most acclaimed restaurants on the island. He uses carefully chosen Corsican produce and his wine list showcases some excellent local vineyards. **anepita.fr**

La Bergerie d'Acciola, Giuncheto
This relaxed yet sophisticated restaurant serves dishes cooked on its wood-fired grill, using top-quality produce. Its wine list gives a full explanation of how each of its chosen wines is made as well as tasting notes. **+33 (0)4 95 77 14 00**

Le Petit Vincent, Bastia
Next to the citadel, this intimate restaurant serves a short but carefully created menu with locally caught fish. Its terrace offers an excellent view of the city's harbour, with blankets on hand for chillier nights. **+33 (0)4 20 00 14 67**

Simon Andrews, A Nepita

'In Corsica's far south, the wine scene offers yet another experience'

Domaine Devichi near Barbaggio

promontory where you can lose yourself in the narrow streets of the old town, visit Domaine Zuria (domainezuria.com). This winery offers tours and tastings (Monday to Saturday 11am-3pm, €24 per person: book ahead) that reveal how the limestone soil in this southern tip offers yet another character for Corsica's wines. And while summer sees the winery host such events as music evenings and hog roasts, autumn is the time when they tend the vineyards with the help of their horse-drawn plough.

Mattei Concept Store

Jardin des Abeilles

THINGS TO DO

Jardin des Abeilles, Ocana
Honey is another of the island's specialities and its flavour varies depending on the season and location of the hives. Tour the bee garden to learn more and try the six different types, including chestnut honey.
lejardindesabeilles.com

Local producers
Visit some of the island's cheese, charcuterie or other food producers, who are signposted throughout the island as part of the Route des Sens Authentiques initiative. They will show you their workshops and you can buy direct. **gustidicorsica.com**

Mattei Concept Store, Bastia
Next to Bastia's main square, Place St-Nicolas, this Art Deco-fronted boutique is owned by the Distillerie Mattei. Its classy, wood-panelled interior has shelves filled with its range of aperitifs, liqueurs and beers, as well as other foods from the island.
distilleriemattei.com/concept-store

Diners relaxing on a restaurant terrace in Bonnieux

LUBERON & VENTOUX

Captivating mountain views, welcoming châteaux and wines that match perfectly with the classic local cuisine. Look ahead to future holidays by planning a road trip through the beautiful countryside that straddles the regions of the Rhône Valley and Provence in southeast France.

As I gaze down the avenue of plane trees towards Château Pesquié, I'm surrounded by much of what the diverse Vaucluse department in France's southeastern country has to offer: verdant vineyards encircle the elegant château, and the white-peaked Mont Ventoux towers above a landscape that cradles the Luberon and Ventoux appellations. Nearby, sleepy Mormoiron is one of the many characterful villages to explore both here and to the south in the Luberon regional nature park. And, as if this wasn't enough to spoil me, the vineyards in the prestigious smaller appellations of Châteauneuf-du-Pape, Gigondas and Vacqueyras are within a 40-minute drive.

At Château Pesquié *(chateaupesquie.com)*, brothers Frédéric and Alexandre Chaudière are the third generation of their family to make wine. They adore the location, with Mont Ventoux lying just to the northeast. 'Here on the Rhône, we're in a corridor between the mountains, but also close to the Mediterranean,' explains Frédéric as he shows me around the vineyard. 'Meanwhile, Mont Ventoux creates a kind of amphitheatre and, although the temperature is around 30°C in the day, it is much cooler at night, which allows for longer maturation – we're one of the last to harvest in the whole area.'

We walk around the organically farmed vineyards, planted with parcels of 50-year-old Carignan vines, along with Cinsault, Grenache and Mourvèdre, Clairette, Roussanne and Viognier, and admire the nine-year-old Syrah vineyards planted with bright yellow blooms of broom and gypsum to improve soil quality.

In the winery, soil samples and detailed three-dimensional maps demonstrate the unique geography and terroir. The Ventoux appellation is quickly growing in popularity and this is one producer ready to offer guests a truly Provençale experience, with hampers for picnics in the gardens, vineyard walks and harvest days.

Nearby, other family-run vineyards such as Domaine du Tix *(domaine-du-tix.com)* share the same 350m elevation, with its hot days and cool nights, and they also welcome visitors for tastings.

Enchanting exploration

To make the most of those hot days, I spend a couple of them in the Parc naturel régional du Luberon, which enjoys a similar climate to the Ventoux appellation just to its north. It's also one of the most enchanting areas in this part of the Provence interior, abounding with golden-stone villages and lavender fields. The terroir offers conditions for vineyards to produce perfect drinks for summer evenings – fruity whites and pale

FACT FILE: LUBERON & VENTOUX
PRIMARY GRAPE VARIETIES

LUBERON

White Grenache Blanc, Clairette Blanche, Vermentino, Bourboulenc, Roussanne, Marsanne, Ugni Blanc, Viognier

Red Syrah, Grenache Noir, Mourvèdre, Carignan, Cinsault

VENTOUX

White Grenache Blanc, Roussanne, Bourboulenc, Clairette, Marsanne, Viognier, Vermentino

Red Carignan, Cinsault, Grenache Noir, Syrah, Mourvèdre

GETTING THERE

The easiest airport to reach is Marseille, which is served by several airlines. The one nearer to the destinations detailed here, however, is Avignon. Both Marseille and Avignon are served by direct trains on Eurostar. From Marseille, it is about a 90-minute drive to Mormoiron; from Avignon, the drive is 45 minutes.

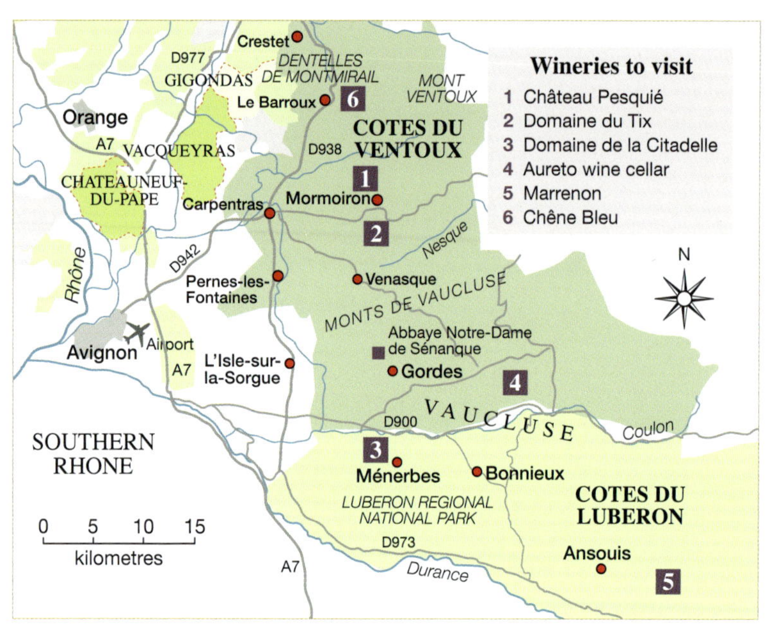

rosés that pair so well with the region's typical fish, olive oil and tomato-based dishes. At the Domaine de la Citadelle *(domaine-citadelle.com)*, near the village of Ménerbes, it isn't only the wide range of excellent wines that draws visitors, but its unusual collection of corkscrews, as well.

The domaine's Musée du Tire-Bouchon holds 1,200 bottle openers of every age, provenance, shape and size. As I gaze into the many cabinets, it strikes me that getting into one's bottle of wine as easily as possible has long been an art form. There is everything from corkscrews that double as shaving brushes, swords or pistols, to some salacious ones – brass legs akimbo, forming a 'T' shape.

Domaine de la Citadelle

'Exploring the different villages is one of the region's greatest pleasures, especially when so many have good vineyards nearby'

On the hill behind the winery and museum, the domaine's botanic garden overlooks the vineyards and Ménerbes perched on a ridge nearby, with raised beds that each contain a different herb or medicinal plant. Inhaling them is as much a treat for the senses as the tasting I take part in when I return to the winery, sipping different blends of the 17 different grapes grown across the vineyard's 50ha.

Exploring the different villages is one of the region's greatest pleasures, especially when so many have good vineyards nearby. After wandering the cobbled streets of Bonnieux, its jumble of terracotta-roofed houses piled up on a hill, I drive on to the Aureto wine cellar *(aureto.fr)*. Belonging to the nearby luxury estate, La Coquillade, the breathtaking views add a further wow factor to the award-winning wines here.

Parc naturel régional du Luberon

MY PERFECT DAY IN VAUCLUSE

MORNING

Take a quick dip in the pool before breakfast at **La Coquillade**, then head out along the road for the 15-minute journey to **Domaine de la Citadelle** (see left), a family-run vineyard near Ménerbes that makes southern Rhône wines from 17 different grape varieties. It also has a quirky corkscrew museum and botanic garden. Enjoy a vertical tasting, then head on towards Gordes, a 20-minute drive north through wild countryside and fragrant maquis. Explore the village, then head north for a hair-raising drive along dramatic gorges to the **Abbaye Notre-Dame de Sénanque** *(senanque.fr)*, where monks grow lavender in the fields outside the famous abbey.

LUNCH & AFTERNOON

Double back for a light but delicious lunch at the bistro restaurant at **Les Bories**, then loop around to the west for a half-hour journey to **L'Isle-sur-la-Sorgue**. This is France's capital of antiques, so spend the afternoon exploring the flea markets on the river's quaysides and the enclaves known as villages for vintage finds and curious objects, as well as high-quality antiques. For a late-afternoon winery tour and tasting, visit **Château Pesquié** near Mormoiron, half an hour north. The winery exhibits give an excellent explanation of the terroirs of the Ventoux appellation, and it offers tastings of the wines made from such varieties as Grenache, Syrah and Viognier.

Abbaye Notre-Dame de Sénanque

EVENING

It's just a 15-minute drive to **Crillon le Brave**, a beautifully designed, 'scattered' hotel that occupies different buildings in a once-dilapidated village. While the gastronomic restaurant has a more sophisticated menu, the view from the terrace of the bistro is second to none: a panorama that encompasses Mont Ventoux's chalky peak, vineyards, olive groves and villages. Tuck into dishes made with fresh Provençale produce as you watch the sun go down.

Château Pesquié looking towards Mont Ventoux

'The terroir offers conditions for vineyards to produce perfect drinks for summer evenings – fruity whites and pale rosés that pair so well with the region's typical dishes'

The architecturally striking winery, with its filigree-iron wall coverings, offers tastings, workshops and tours. Further to the southeast, beyond Ansouis, a village with a medieval château, you can explore the vineyard of the Marrenon wine estate *(marrenon.com)* in the company of one of its winemakers, complete with a gourmet picnic among the vines.

No boundaries

I finish my journey with a foray back to the north, where a patchwork of appellations is cradled in the southern Rhône valley. To the west, below Orange, are the exclusive vineyards of Châteauneuf-du-Pape, and I drive past those of Côtes du Rhône cru villages Beaumes-de-Venises, Vacqueyras, Gigondas and Séguret in quick succession. Where I'm destined, though, is the crossroads at which four of the southern Rhône's appellations meet.

On the hill behind the village of Crestet, I drive narrow, pine-shaded lanes to find Chêne Bleu *(chenebleu.com)*, a wine estate that envelops you in its natural surroundings. After buying it in 1993, Xavier and Nicole Rolet painstakingly restored the medieval priory at its heart, seeing the potential of the vineyard with an altitude of 550m. As I arrive, Xavier's daughter Danielle is there to show me around the extraordinary setting, which is overlooked by the Dentelles de Montmirail. As I admire the saddle of land where the Ventoux, Séguret, Côtes du Rhône and Gigondas appellations meet, I hear a cuckoo call, frogs croaking and the warm breeze blowing through the trees. Here they welcome the Mistral wind; it helps fend off disease in the vines.

We continue inside, where Danielle points out the level of dedication and adoration that has gone into the winery; from Nicole's exquisite friezes that adorn the walls surrounding the concrete vats, to humorous medieval-style wine labels. Afterwards, I taste the acclaimed wines that do not adhere to appellation boundaries. 'It was a brave move as first-time winemakers, but it's paid off!' says Danielle.

When I leave, she thanks me for making the tricky journey to see them, but with chambres d'hôtes, gourmet dinners and wine courses on offer, it's a trip I'm willing to make again.

YOUR LUBERON & VENTOUX ADDRESS BOOK

ACCOMMODATION

Château de Montcaud, Sabran
A short hop over the Rhône river, this new four-star hotel run by friendly Swiss couple Rolf and Andrea Bertschi has a superb restaurant and a very knowledgeable sommelier. Makes a good base if you're including Châteauneuf-du-Pape in your trip.
chateaudemontcaud.com

Crillon le Brave
This 'scattered hotel', which occupies historic buildings in the village of the same name, enjoys a breathtaking view of Mont Ventoux. Its refurbished rooms were designed by top French architect Charles Zana. **crillonlebrave.com**

La Coquillade, Gargas
An eco-friendly and luxurious resort looking out over the surrounding vineyards belonging to its own winery Aureto. Explore the area on high-quality bicycles from the on-site cycling centre. **coquillade.fr**

RESTAURANTS

La Figuière, Fontaine de Vaucluse
Tucked away in a corner of the village near the river Sorgue, this classic bistro restaurant offers an idyllic courtyard under the shade of parasols. Its menu features French and Provençale classics. **lafiguiere-provence.fr**

La Table de Xavier Mathieu, Joucas
In a small village between Gordes and Roussillon, chef and Provence native Xavier Mathieu brings together Provençale ingredients for his exquisite dishes, served on the terrace of the restaurant, overlooking the Luberon valley. **lephebus.com**

Les Bories, Gordes
This one-star Michelin restaurant near Gordes sees chef Grégory Mirer, who previously worked for the late, celebrated Joël Robuchon in Paris, serve refined dishes using the best Provençale produce. **hotellesbories.com**

SHOPS & LEISURE

Les Délices du Luberon
Stock up on delicious produce to take home with you at Les Délices du Luberon, where you'll find jars of tapenade, traditional Provençale herbs and delicious olive oils. Shops in L'Isle-sur-la-Sorgue, Avignon and St-Rémy-de-Provence.
delices-du-luberon.fr

L'île aux Brocantes
The town of L'Isle-sur-la-Sorgue is the French capital of antiques and bric-a-brac, where you'll find chic stores alongside its canals, as well as brocante villages such as L'Ile aux Brocantes, where several dealers assemble in each space.
lileauxbrocantes.com

Sun-E-Bike
The Luberon region is hilly, but with the help of an electric bike, a pedal between vineyards can be a pleasure rather than a challenge. This bike rental service offers self-guided vineyard tours starting from Bonnieux (as well as other locations).
sun-e-bike.com

Les Bories

Les Délices du Luberon

108
TAXIS

REIMS

The effervescent attractions of France's Champagne region require little introduction, and the city of Reims provides the perfect base, offering history, culture – and a very smart food scene.

Next time you plan a wine trip to Champagne – stocking up for the festive season, perhaps? – make Reims your destination. Culturally jewelled and easy to reach, it just pips Epernay, Champagne's other wine town, which has a rather straitlaced vibe in comparison.

The city has long ago outgrown its days as a one-night stopover by car en route to Burgundy or southern French sun. It's now a destination in its own right for Champagne fans on long weekend breaks. The major Champagne houses are based here, and it has a very smart food scene. Its restaurants total eight Michelin stars – nine if you include Royal Champagne.

Long-running city-centre building works have resulted in a smart new tram system, and its cathedral, the basilica and former abbey of St-Remi (now home to a museum), and the surrounding area's vineyards are already UNESCO Heritage Sites. This compact modern city, full of young professionals and with an international student cohort, sprouts new architecture alongside the old and new businesses by the day and has a cool, hip buzz. Its bid to be a European Capital of Culture 2028 is just one demonstration of that. The atmosphere is almost Parisian, but with more Champagne.

First-time visitors to Reims, why not join one or two Champagne houses' paid tours? Mumm, Pommery and Veuve Clicquot are slickly informative, but Charles Heidsieck, Ruinart and Taittinger are my tip, with their jaw-dropping, soaring chalk cellars. For a more personal experience, visit small family domaines just 30 minutes from Reims centre by car or taxi. The village of Ecueil is a hotbed of creativity, home to elite grower Champagnes Lacourte-Godbillon, Nicolas Maillart and Savart, and Emmanuel Brochet in nearby Villers-aux-Noeuds. Or you could try your luck at getting a visit to the superb Champagne Bérêche in Le Craon de Ludes.

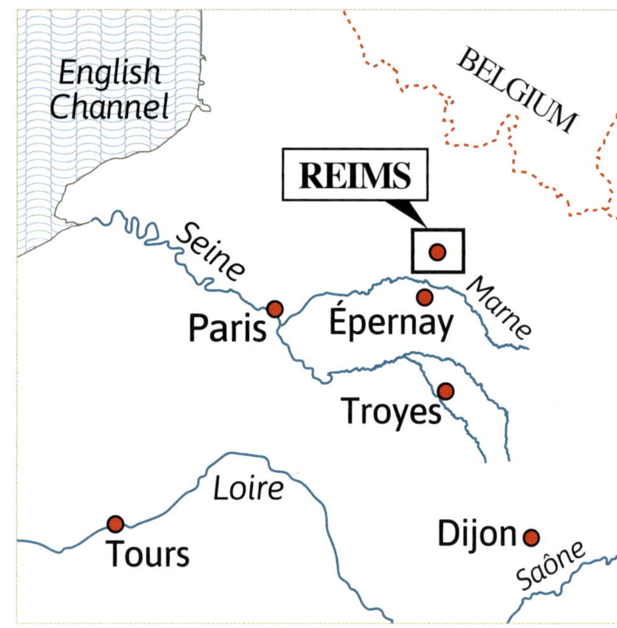

GETTING THERE

If you're travelling from the UK, Reims is one of few French wine cities reachable before lunch. It's only 47 minutes by fast train from Paris Gare de l'Est, which is a short walk from the Eurostar terminal, if you started in London. By car, Reims is approximately six hours from London.

If you're flying in from further afield, and clear arrivals quickly at Paris Charles de Gaulle airport, you can be sipping Champagne at your hotel bar in Reims just 90 minutes after landing, via train or car. If carbon guilt and expense is no issue, fly privately into Reims-Prunay airport, just to the southeast of the city.

For getting around once you've arrived, use private chauffeur firm HM Prestige VTC, which also operates airport shuttles. **hm-prestige-vtc.fr**

For guided tours at the major Champagne houses, you'll need to book ahead in the spring and summer (when most people visit). If you want to visit small growers, remember they won't have a dedicated visitor team; they tend the vines and make the wine so you must email or phone well ahead and be there on time. For all that, the welcome will be warm and you will learn a lot.

YOUR REIMS ADDRESS BOOK

ACCOMMODATION

Hôtel de la Paix Popular, central, well-run, and a cut above most Best Westerns. **9 Rue Buirette +33 (0)3 26 40 04 08**

La Caserne Chanzy
New hotel with some heavenly rooms offering a view of the cathedral.
18 Rue Tronsson Ducoudray lacasernechanzy.com

Novotel Suites Reims Centre I often stay here as it's convenient for the train station and near the smart Boulingrin quarter, whose glory is the refurbished **Halles Boulingrin** food market. Big rooms and great as a workspace, with famous Novotel attentiveness and easy parking.
1 Rue Edouard Mignot all.accor.com

Résidence Eisenhower For somewhere more luxurious, this sumptuous mini-palace is Reims' newest eyrie.
17 Bd Lundy residence-eisenhower.com

And, of course, there's always **Airbnb**.

RESTAURANTS & CAFES

Au Bon Manger Reims' coolest hangout for organic and biodynamic Champagne, and a chance to rub shoulders with some of the region's rockstar growers. The mainly cold-plate food is from top suppliers.
aubonmanger.fr

Au Cul de Poule Hearty food but the cooking is of real skill and there's a joyous, unpressured ambience. I love the steak tartare, prepared any of seven ways. A great Champagne list, with big houses and growers alike.
auculdepoule.com

Domaine Les Crayères A superlative hotel and two-star Michelin restaurant with tranquillity, special service and Champagne in depth. Sip on the terrace, overlooking Champagne's largest lawn. **lescrayeres.com**

L'Assiette Champenoise Simply the best. Unremittingly high-end it is, with its three Michelin stars, but it's not stiff. And there's stylish accommodation to be enjoyed, too.
assiettechampenoise.com

La Brasserie Le Boulingrin Over the years I've probably eaten more oysters and downed more steak frites and cod mornay here than anywhere else. **boulingrin.fr**

Le Bocal Strictly for fish fanatics, and in the lively and fashionable Boulingrin district. Select a live lobster from the tank or enjoy oysters with their greatest accompaniment – Champagne!
restaurantlebocal.fr

Racine

Le Coq Rouge Popular bar à vins where you're likely to find Reims' professional jeunesse dorée rubbing shoulders with Champagne makers, enjoying great bistro food and an extensive Champagne list. If nothing else, order the gambas en tempura – big prawns with a light batter and spicy dip. Fabulous with Champagne Diebolt-Vallois, Fleur de Passion 2012. The sister and brother places, **Le P'tit Coq** and **La Braise**, are good too. Champagnes at prices close to UK retail. **facebook.com/lecoqrouge.reims**

Le Wine Bar by Le Vintage Maybe the best Champagne list anywhere – small-plate food, but the action is in that list. Get there early in spring and summer. Perfect for group taste-athons. **winebar-reims.com**

Racine Reims' smash-hit Japanese-French fusion restaurant holds two Michelin stars and is run by chef Kazuyuki Tanaka. The city's hottest restaurant ticket right now: book well ahead. **racine.re**

Royal Champagne Recently refurbished, this is now Champagne's smartest hotel. Its main restaurant quickly grabbed a Michelin star and the bistro Le Bellevue makes delicious food, too – and if the sun allows, you will have the best terrace meal and view in all Champagne. It's in Champillon, which is more than halfway to Epernay, so it's a taxi ride out.
royalchampagne.com

The Glue Pot It's best to avoid the often-heaving Place Drouet d'Erlon, a tourist haven of hotels and bars, but make an exception for The Glue Pot, which serves excellent grower Champagnes in Zalto glasses. The food is so-so (although a friend loves the steak tartare here!), but there's a fun vibe. Good for leisurely tasting outside mealtimes with a group.
thegluepot.com

Royal Champagne, Champillon

TOURS & TOURAINE

Opulent châteaux, manicured gardens, relaxation on or by the river, and a fantastically diverse wine scene on tap – yet this enchanting Loire valley region still flies comparatively low on holidaymakers' radar.

Rolling hills in Burgundy; the sun-baked slopes of Provence; the grand estates of Bordeaux – they all evoke a clear holiday picture. In comparison, lesser-known Touraine is trickier to pin down in the mind's eye. It takes a real-life visit to this underrated appellation in the Middle Loire to realise the abundance of its appeal. But in truth, the 'Garden of France' has everything you could want in a Gallic getaway.

There's the endless series of opulent châteaux, filled with treasures and flanked with formal gardens clipped by box hedges. A handsome regional capital, Tours, with half-timbered architecture and Joan of Arc history. Sleepy little villages with cafes turning out salads topped with sharp local goat's cheese; lush, trail-lined Loches forest. And, of course, rippling through all of it, the mighty Loire river, the region's lifeline. Once a key trading route, for you it equals leisurely evening cruises, afternoon kayak sessions or scenic waterside cycling routes.

And, of course, there is the wine. Despite a surface area of just 5,500ha – stretching roughly 100km from around Blois in the east to Bourgueil in the west – the Touraine appellation claims

Château d'Amboise (on the right), and the Pont du Maréchal Leclerc across the Loire river and the island of Ile d'Or

'The Garden of France has everything you could want in a Gallic getaway'

vast stylistic diversity. There are lively sparklings and luscious sweet whites; juicy rosés and generous reds. Celebrated names such as Chinon and Vouvray sit alongside little-known ones such as Montlouis-sur-Loire (Vouvray's reflection just across on the south side of the Loire river).

Most producers are small, often working across multiple styles and sub-regions. Humble tasting rooms ooze familial warmth, with many set atop troglodyte cellars – resulting from tuffeau limestone being harvested to build the region's magnificent châteaux. Flinty clay (perruches), sand and gravel soils realise Chenin Blanc and Cabernet Franc – the Loire poster grapes, white and red – in the full spectrum of their diversity. Rarer Romorantin, Arbois and Pineau d'Aunis join the mix, too, alongside the likes of Sauvignon Blanc, Grolleau Noir and Côt (Malbec). Whichever cellar door you end up at, one thing Touraine tastings are not is boring.

Each day holds a new adventure, always backlit by resplendent architecture. Begin in gateway city Tours, in the heart of the region, about a 90-minute direct train ride from Paris, Gare Montparnasse. Here, a magnificent cathedral is testament to

'Whichever cellar door you end up at, one thing Touraine tastings are not is boring'

the city's medieval grandeur, while creaking buildings once hosted the likes of Joan of Arc. Visit the photogenic old centre, then pop into little wine bars pouring bargainous tipples (sometimes £3 a glass) to get a vinous overview of the region. Pork rillettes or piquant local goat's cheeses such as ashy Ste-Maure de Touraine provide the perfect stomach lining.

An abundance to savour

Then, strike out. Barely outside the city walls to the east, sub-region Vouvray produces one of Touraine's best-loved wines from Chenin Blanc on a plateau hugging the Loire river. At acclaimed Domaine Huet *(domainehuet.com)*, a local standard-bearer since 1928, you can see all Chenin is capable of, from sparkling to still, sec to moelleux (gently sweet). Just a minute down the road, the 20ha Le Clos de L'Epinay (vinvouvray.com) is family-owned and friendly. Meanwhile, 10 minutes northeast of Le Clos, at Château de Valmer *(chateaudevalmer.com)* the formal gardens are as appealing as the award-winning brut sparkling and sec. Winemaker-owner Jean de Saint Venant also makes a solid AP Touraine rosé sparkling from 100% Grolleau, and it costs just £8 a bottle.

On the other end of the spectrum, about 50km southwest of Tours, lies Chinon – heartland of Cabernet Franc. The UNESCO-listed central village on Loire tributary the Vienne is picturesque; crowned by a crumbling medieval royal fortress and cluttered with bistros for epic weekend lunches. The fabled red wines from the surrounding vineyards are equally wow-factor. Tick off Bernard Baudry *(bernardbaudry.com)*, which produces organic wines from old vines rooted in a wide range of soil types; try concentrated Les Grézeaux from a 50-year-old plot. Respected domaine Charles Joguet *(charlesjoguet.com)*, one of the region's pioneers in single-vineyard bottlings, is another must. It produces nearly a dozen

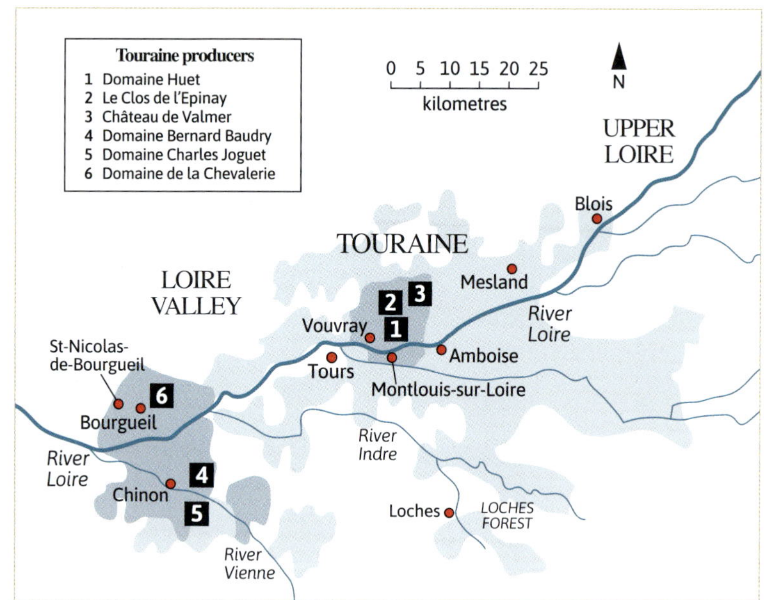

GETTING THERE

A fast train into Tours takes about 90 minutes from Paris, accessible from the UK via Eurostar (eurostar.com). To explore you need a car; it takes about 45 minutes to drive from central Tours to the region's furthest reaches east or west.

MY PERFECT WEEKEND IN TOURS & TOURAINE

FRIDAY

Kick off with a history lesson in gateway city Tours, taking in the cathedral on Rue Lavoisier and the half-timbered architecture on Place Plumereau (above). Then strike out to pretty riverside Amboise – 30 minutes east – home to a must-see château and cute restaurants. Get an overview of Touraine wines on Place Michel Debré with tastings at **La Cave** *(@lacaveamboise)*, and riverside cellars **Caves Ambacia** *(caves-ambacia.fr)* on Rue du Rocher des Violettes. The latter sells Vouvray demi-sec vintages back to 1874 and 1990s Bourgueil in magnum, with lunches of house-aged goat's cheese and charcuterie. Afterwards, take a boat ride on the Loire or sample the Loire à Vélo cycling route to nearby Chaumont-sur-Loire. See the art-filled château and gardens then stay over in **Le Bois des Chambres**.

Caves Ambacia

SATURDAY

Today it's all about Cabernet Franc: Chinon and Bourgueil. Start in the former, hitting esteemed **Bernard Baudry**, then **Domaine Pierre et Bertrand Couly** *(pb-couly.com)* for a Segway tour through the vines. Have lunch in pretty Chinon by the square at Au Chapeau Rouge, then burn it off walking around the **Royal Fortress of Chinon** *(forteressechinon.fr)* ruins. Make the half-hour drive to Bourgueil to visit the stunning troglodyte cellars of biodynamic **Domaine de la Chevalerie**, run by young siblings, then stop off at the epic gardens of **Château de Villandry** on your way back towards Tours for the night. Check into Ferdinand Hotel and hit a local wine bar for dinner.

SUNDAY

Make your way southeast to **Château de Nitray** *(chateau-nitray.fr)* for a lively range of Touraine wines made from Côt, Grolleau and Sauvignon. If they're serving up steaks for lunch in their on-site restaurant (call ahead to check), have lunch here, then head onto **Château de Valmer**, where more sublime landscaped gardens and a troglodyte chapel come with award-winning Vouvray. Finally, check into **Loire Valley Lodges** *(loirevalleylodges.com –pictured, top)* mid-afternoon to make the most of its lovely pool, vegetable gardens and cute hens. Have dinner in the restaurant before bedding down in one of the photogenic luxury treehouses, decorated by international artists.

Place Plumereau in Tours old town

Loire Valley Lodges

Château de Nitray

different cuvées, from fruity Les Petites Roches to powerful Clos de la Dioterie. Afterwards, nip to nearby AP Bourgueil. The region's Cabernet Francs don't yet have the same following as Chinon's, but there are real treasures to be found. Domaine de la Chevalerie *(domainedelachevalerie.fr)* ages biodynamically grown wine in its troglodyte cellars, including the lively, leathery Bretêche wrought from calcareous clay soils.

So many options

Vouvray and Chinon may be the headliners, but in between there's much more in Touraine to keep your tastebuds entertained. In Touraine Amboise, try everything from Cabernet Sauvignon to Arbois. Touraine Mesland, meanwhile, serves up Côt and lighter-bodied Gamay. The more time you take to explore, the more delicious discoveries you will make.

In between tastings, you won't be short of things to do – arguably much more than in most wine regions. Cycle a stretch of the 900km Loire à Vélo cycling route (see box, below right), zipping through forest, vineyards or along riverbanks (electric bikes are available to hire in many towns). Hop on a traditional wooden merchant boat for a slow slink down the river at sunset *(momentsdeloire.fr)*. Or tour the many châteaux, from UNESCO-listed Amboise *(chateau-amboise.com)* – final resting place of Leonardo da Vinci – to contemporary art-stuffed Domaine de Chaumont-sur-Loire *(domaine-chaumont.fr)*, which also has exceptional gardens.

And at the end of a long day exploring, you can check into a lovely hotel; there are increasing numbers of stylish options both in Touraine itself and just outside its borders. The past couple of years alone have seen the launch of a sister hotel to Bordeaux's Les Sources de Caudalie, Les Sources de Cheverny *(sources-cheverny.com)*, and Fleur de Loire *(fleurdeloire.com)* in Blois, both of which have attracted a cool new high-end clientele. And that means one thing: just like Burgundy, Provence and Bordeaux, it's only a matter of time until this underrated pocket of the Loire valley is truly discovered.

YOUR TOURS & TOURAINE ADDRESS BOOK

ACCOMMODATION

Château de Pray, Chargé
If it has to be a château stay, this Renaissance pile, near Amboise on the south bank of the Loire river, comes with 5ha of wooded parkland, a one-star Michelin restaurant, heated pool and rooms dripping in Toile de Jouy. **chateaudepray.fr**

Ferdinand Hotel, Tours
This 14-room stay in the heart of the shopping district is one of Tours' more chic places to stay. Cosy rooms come dressed in loud wallpapers and minimalist furniture. **ferdinandhotel.fr**

Le Bois des Chambres, Chaumont-sur-Loire
In the east of the region, this recent launch on the grounds of historic Domaine de Chaumont-sur-Loire shows that wine region accommodation isn't all stuffy old beamed ceilings. Barns have been converted into sleek, pastel-hued eco-friendly rooms with views out to a central courtyard garden. **leboisdeschambres.fr**

FOOD & DRINK

Ardent, Esvres
Chef Gaëtan Evrard's restaurant at chic hotel Loire Valley Lodges shows off the contemporary side to Touraine cookery, with a menu built around produce from the hotel's veg garden. Dinner is wow-factor but lunch has the benefit of what may be your best-ever croque monsieur: thick farmhouse bread with mushrooms, ham and Mornay sauce with aged Comté. **loirevalleylodges.com**

Au Chapeau Rouge, Chinon
It doesn't get more Gallic than this red awning-fronted bistro on Chinon's main square, in the shadow of the UNESCO town's historic fortress. Lazy, wine-soaked lunches – think gilt-head bream with white butter and Touraine saffron – spill out onto the cobbled street. **auchapeaurouge.fr**

Tutu, Tours
You'll get a wine tour around Touraine at this forward-thinking, funky wine bar in Tours. Glasses start at €3.50 (£3.00) and pours range from a 1996 Montlouis sur Loire by Clos Habert to a 2017 orange wine from Laurent Lebled. After oysters and charcuterie, order local favourite tarte vigneronne (apple tart with wine jam). **tutu-tours.eatbu.com**

hâteau de Villandry
see box, below right)

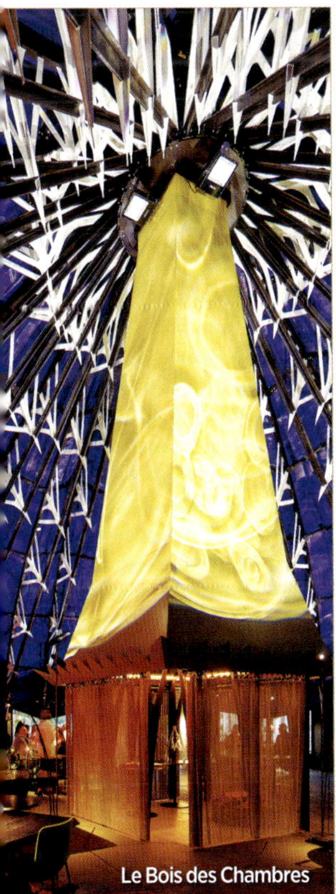
Le Bois des Chambres

Au Chapeau Rouge

Ardent

Tours cathedral

THINGS TO DO

Château de Villandry, Villandry

If the Loire valley is famous for châteaux and gardens, this pile – once home to Napoleon Bonaparte's brother – shows off the best of both. Visit the elaborate rooms, then explore the vast formal gardens with more than 50km of box plant hedging and 1,000 lime trees. **chateauvillandry.fr**

Loire à Vélo

This 900km cycling route follows the mighty Loire river, cutting through vineyards, forest and farmland. Several local operators rent electric bikes; do a multi-day stretch from Amboise to Chinon via Tours, or just a lazy afternoon, stopping off at wineries as you go. **loireavelo.fr**

Saint-Gatien Cathedral, Tours

Ornate archways, intricate stained glass, flamboyant flying buttresses – Tours' top architectural masterpiece is the Loire's answer to Notre-Dame. It looks even better at night, when golden lighting makes the stone facade details really pop. **Rue Lavoisier/Rue Fleury**

ITALY

Drive through Tuscan vineyards dotted with cypress trees, sip on crisp whites in Alto Adige while you take in mountain views, or dine on bright yellow fresh pasta covered in shavings of fresh truffle, a glass of Nebbiolo at hand, in Piedmont. Just a few of the delights that await in a country famous for its fascinating indigenous grape varieties.

ALTO ADIGE

In Italy's far north, this is a place that's ripe for exploring, home to dramatic Alpine scenery and superb skiing, Mediterranean vegetation and sweltering summers – and wines that blend Germanic precision with Italian flair.

Matched in excellence by quality cuisine, the wine culture in Alto Adige (also known as South Tyrol) is deep-rooted. Vines grow at varied altitudes in a Y-shaped area, with the main city, Bolzano, at its heart. Terrains here range from sandy marl and limestone to schist or porphyry, and wide temperature swings make for wines that combine the fresh, fragrant aromas of cooler climates with warm-weather power and complexity. Most come under the six Alto Adige DOC sub-zones, while more specific MGA zones (menzione geografica aggiuntiva: a specific, delimited area within a DOC/G, broadly equivalent to a 'climat' in France) are planned.

Bolzano's appealing historic centre has countless references to wine in road names, frescoes and the cathedral's medieval 'wine door'. Protected by mountains, the city registers some of Italy's hottest summer temperatures and vines flourish here. These include the Lagrein vines surrounding Castel Mareccio (see 'My perfect day', p65). The true home of fruity, vibrant purple Lagrein is, however, across the Talavera river in the Gries neighbourhood. At the Muri-Gries monastery (*muri-gries.com*), a former fortress that's been home to Benedictines since 1845, benchmark Lagreins include Klosteranger, a single-vineyard Riserva from vines grown within the monastery walls.

Another monastery that plays a key part in Alto Adige winemaking is Abbazia di Novacella (*kloster-neustift.it*). Active since 1142, it's among the world's oldest wineries and is fascinating to visit. Here in the cool Isarco valley, northeast of

FACT FILE: ALTO ADIGE

PLANTED AREA 45,600ha

ALTITUDE RANGE 200m-1,000m

DAYS OF SUNSHINE 300 per annum

ANNUAL PRODUCTION 40 million bottles (64% White, 36% Red); 98% are DOC

SOURCE: CONSORZIO VINI ALTO ADIGE, 2021 ALTOADIGEWINES.COM

THIS PAGE AND OPPOSITE Abbazia di Novacella

Bolzano, white grapes such as the semi-aromatic Kerner thrive; Sylvaner, the winery's speciality, also grows well here – a new sparkling version is due for launch in December 2022. The producer is also one of several experimenting with fungus-resistant [hybrid] PIWI varieties.

Also in the Isarco valley, at the 3ha Röck estate *(roeck.bz)*, Hannes Augschöll is among the new wave of winemakers bringing changes; his range of low-intervention wines includes a Müller-Thurgau pét-nat. 'I'm aiming for drier, lighter wines than my father's,' he says. Each autumn, the winery, along with many others, opens its doors for the Törggelen festival, serving home-cooked dishes and speck (cured ham) that's been smoked at the stone farmhouse.

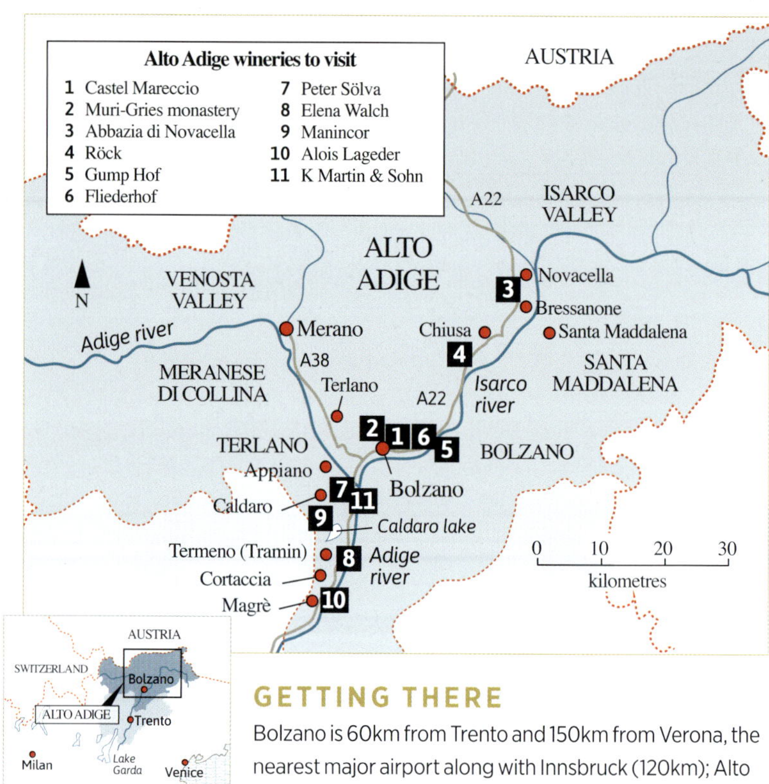

GETTING THERE

Bolzano is 60km from Trento and 150km from Verona, the nearest major airport along with Innsbruck (120km); Alto Adige Bus (altoadigebus.com) runs regular shuttles. Bolzano airport has some international flights.

At the southern tip of the Isarco valley, with steep 70° slopes, Gump Hof *(gumphof.it)* has been run by the Prackwieser family for two centuries. Markus, in charge since 2000, introduced Sauvignon Blanc here, and his wines have an irresistible intensity. He's one of many planting at ever-higher altitudes; 2021 was the first harvest at 800m, near the striking Presule Castle.

Schiava Revival

Another young winemaker bringing new ideas is Martin Ramoser (pictured right) at his family's Fliederhof winery in Santa Maddalena di Sotto *(fliederhof.it)*, where pergola-trained Schiava vines surround the pretty hillside church, walkable from Bolzano via the St-Oswald footpath. Here, a percentage of inky Lagrein is typically added to naturally pale Schiava (also known as Vernatsch). After dominating the region's winemaking for centuries, the fresh, light variety has endured a generic table wine image, but things are looking up. 'I really believe in Schiava's renaissance,' says Ramoser. 'It's so versatile, and ideal with contemporary cuisine and our local specialities.'

At Fliederhof winery, from right: Martin Ramoser, his parents Stefan and Astrid, and his wife Katja

Schiava is also the main grape in the Merano sub-zone, northwest of Bolzano, and the Caldaro DOC area around

Caldaro lake to the south. With neighbouring Appiano, Caldaro is the real heart of the region's wine scene and central to the Strada del Vino *(suedtiroler-weinstrasse.it)* with its network of paths through vineyards. Regular events at Caldaro include the 'Fascination wine world South Tyrol' sommelier talks and tastings, held on Sundays from April to November; the Kalterer Weintage 'wine days' in the market square on 1-2 September; and Weinkulinarium ('Wine & Cuisine') on Saturday 1 October *(see kaltern.com)*. It is home to a historic wine museum and wineries such as Peter Sölva *(soelva.com)*, run by the same family for 10 generations making some memorable wines including a Merlot-Cabernet Sauvignon-Lagrein blend, Amistar.

'Schiava is so versatile, ideal with contemporary cuisine and our local specialities'

Martin Ramoser (below, far right)

Piazza Erbe market, Bolzano

MY PERFECT DAY IN ALTO ADIGE

MORNING

Start the day at the Piazza Erbe market in central Bolzano to buy locally grown apples, then pick up spicy pastries or choose from a wide range of breads at the **Franziskaner bakery** *(@franziskanerbaeckerei)* around the corner. Pop into the Franciscan church next door to see medieval frescoes before taking your purchases for a breakfast picnic at the Talavera riverside park, stopping for coffee on the way. Take a short stroll to see the vines at **Castel Mareccio** *(maretsch.info)*, then cross the river into the Gries neighbourhood. Visit the imposing Muri-Gries monastery and try a selection of wines, including some of the very best Lagrein there is. Before leaving the area, walk through exotic vegetation along the zig-zagging Guncina footpath for a stunning panoramic view across the city to the Dolomites.

LUNCH & AFTERNOON

Pick up your car and head southeast and up another hillside for lunch at the idyllic **Grafhof*** farmhouse, just 5km out of town – though it feels much further than that. Tuck into a meal of farm-fresh eggs with potatoes and speck, followed by strudel made with home-grown apples, all while enjoying the fabulous views across the valley to the Santa Maddalena hill and its tapestry of pergola-trained vines. After lunch, drive south along the Strada del Vino to Caldaro. Explore the attractive town centre, visit the **Museo del Vino** *(weinmuseum.it)* and stop at the Peter Sölva winery. Walk through vines on the Sentiero del Vino footpath to see Caldaro lake, ending up back in the centre at the **Casa del Vino Punkt*** wine bar for a personalised tasting of local wines.

EVENING

From Caldaro it's just 7km to **Hotel Weinegg***. After checking in, cross the hotel garden and vines beyond to the **K Martini & Sohn** winery *(martini-sohn.it)*. Try some of its excellent wines or relax at a barrel table on the patio with a glass of old-vine Pinot Bianco. Enjoy a delicious dinner on the hotel terrace – perhaps local char with white wine mousse and something from the extensive wine list – then finish the day with a relaxing swim under the stars in the hotel's heated indoor-outdoor pool.

For details of entries marked with an asterisk (*), see p67

Alois Lageder

With so many growers owning such tiny plots (on average a single hectare), Alto Adige's admirably run cooperatives are fundamental to the region's wine production. But rather than churning out lowest-common-denominator wines, many are impressively experimental. The Caldaro co-op is currently trying out different Schiava clones, while Cantina Girlan *(girlan.it)* has launched a Pinot Nero project, and St Pauls (stpauls.wine) is working with Pinot Bianco, the variety grown throughout the region that's a locals' favourite for its dry, peppery bite. It's also a key player (with Pinot Nero and Chardonnay) in the growing movement for Alto Adige traditional-method spumante.

Idyllic Setting

The landmark 17th-century Castel Ringberg has fabulous views from the terrace across a patchwork of vines to Caldaro lake. Owned by the Elena Walch winery *(elenawalch.com)*, a protagonist in the recent drive for quality, it's run with style and brio by Walch's two daughters. The main site at Termeno (Tramin in German), the village Gewürztraminer is named after, has a historic cellar with traditional sculpted barrels. While locals are fond of recalling that Gewürztraminer isn't popular here, some excellent versions of the aromatic variety are produced.

Nearby at Manincor *(manincor.com)*, Count Michael Goëss-Enzenberg and his wife Sophie are hands-on at their biodynamic estate, in an idyllic setting with sheep and chickens among the vines, a stunning manor house dating from 1608 and a modern underground cellar; no wonder it was chosen as the setting for the 2021 TV mini-series The Winemaker (Good Friends Filmproduktion). Barrels are made from estate-grown wood and the wines, including the Mason Pinot Nero, share an elegance that reflects the location.

Horses and oxen are among the animals grazing the vineyards at the Alois Lageder winery *(aloislageder.eu)*, a collection of atmospheric farm buildings in Magrè at the southern end of the Strada del Vino. Many of the wines have Demeter biodynamic certification, and the top range, which ages to the sounds of Bach's Brandenburg Concertos, includes an exceptional Chardonnay and a historic field blend of Cabernet Sauvignon, Carménère, Cabernet Franc and Merlot from 140-year-old vines at the Löwengang estate. But they're mere youngsters compared to a vine planted in the village in 1601, a natural monument that's still producing grapes today.

Alto Adige is the diamond tip of the Italian wine scene and much of its wine never leaves the region – all the more reason to plan a trip to this extraordinary wine-lover's wonderland.

YOUR ALTO ADIGE ADDRESS BOOK

ACCOMMODATION

Angerburg, San Michele Appiano
This simple yet comfortable, family-run hotel stands in a large garden with an orchard and pool. Balconies have spectacular views and the restaurant serves traditional dishes and local wines. **hotel-angerburg.com**

Hotel Weinegg, Cornaiano
A five-star resort hotel with a well-equipped spa and indoor-outdoor pool, plus superb dining and an impressive wine list including bottles from the owner's family estate. **weinegg.com**

Panholzer, Caldaro
Stay at the suite among the vines near lake Caldaro and tuck into beautifully presented seasonal dishes at the gorgeous stone farmhouse restaurant with courtyard garden. **panholzer.it**

Löwengrube

RESTAURANTS AND BARS

Alois Lageder Paradeis, Magrè
Named after the mountain behind, this organic-certified winery restaurant specialises in light and tasty dishes with much use of estate-grown vegetables, paired, of course, with Alois Lageder wines. **aloislageder.eu**

Casa del Vino Punkt, Caldaro
Both a locals' bar and showcase for Caldaro wines, this informal 'house of wine' on the central piazza offers 20-25 wines by the glass, personalised tastings and typical snacks including smoked sausage and local cheese. **@weinhauspunkt**

Grafhof, Bolzano
The Staffer family's historic buschenschank (farmhouse winery serving food), which overlooks Bolzano, has a wood-clad stube and terrace. Enjoy tasty home cooking such as spinach-filled pasta, paired with Schiava wine. **buschenschankgrafhof.it**

Löwengrube, Bolzano
Tradition and design combine at Bolzano's oldest restaurant, dating from 1543. There are medieval frescoes, a thousand-strong wine list and a menu featuring dishes such as braised lamb cannelloni. **loewengrube.it**

Rifugio Oberholz, Obereggen
Take the Obereggen chairlift or walk (90 minutes) to this contemporary mountain hut (2,096m above sea level) for spectacular scenery and superb cuisine. Try the Alpine beef with a Lagrein Riserva and sweet kaiserschmarrn pancakes. **oberholz.com**

Vögele, Bolzano
A local institution with an all-day menu of traditional specialities, this historic tavern guarantees at least 75% regional wines as part of the Locanda Sudtirolese association. **voegele.it**

COMO

Long an exclusive destination, this delightful waterside city offers the best of Italian style, accommodation and indulgent days on the water, all surrounded by stunning mountain scenery.

Less than an hour north of Milan and just 6km from the Swiss border, the charming city of Como sits at the southern tip of its eponymous, inverted Y-shaped lake's western leg, in a narrow bay flanked by wooded morainic hills. This breathtaking natural landscape has been used as a movie location countless times, and many actors and celebrities own villas here: George Clooney and Brad Pitt, among others.

Traces of viticulture at Lake Como date back to the Rhaetians in the Iron Age, and Leo Tolstoy mentioned it in the 19th century. But vines were abandoned in favour of mulberry cultivation for silk manufacturing – Como is famed for its silk, and its Museo della Seta silk museum (*museosetacomo.com*) is worth a visit.

Como's recent winemaking revival highlights the characteristic features of a sunny, windy terroir with its own special microclimate. Its savoury, light-bodied wines pair beautifully with classic lake-fish dishes such as perch risotto or grilled whitefish.

Nestled within the city walls are Como's genteel porticoes, set among Renaissance houses and the fine Duomo. The Tempio Voltiano (*alessandrovolta.it*) is the museum dedicated to physicist Alessandro Volta, the illustrious Como-born inventor of the electric battery.

But to really feast your eyes, take a romantic stroll along the lakeshore, admiring villa after villa and their beautiful gardens. Also make sure to visit the neo-classical gem Villa Olmo, which hosts exhibitions and events. Wander back to the city centre via the pier that leads to the Life Electric sculpture and stand surrounded by water, admiring the lake in all its glory – sunset is the ideal time.

To truly understand the local saying, 'the lake should be seen from the lake', meander through its least explored corners towards the only island, Comacina (where spectacular firework displays mark the feast of San Giovanni in late June); or hop over to the 'pearl of the lake', Bellagio, on the tip of the promontory where the lake splits into two legs. The shoreline is dotted with beautiful mansions, such as Villa Erba in Cernobbio and Villa Carlotta (Tremezzo), as well as tiny Romanesque churches and medieval villages. So without further ado, rent a boat from Tasell (*tasell.com*), which has been in the business for more than a century.

Not to be missed is the Como to Brunate funicular: the panorama from the top is unforgettable. Walk through the village, famous for its art nouveau villas, up to the Voltiano lighthouse, and climb to the top to be rewarded with some awe-inspiring views!

LEFT A glimpse of Lake Como from one of the picturesque streets of Bellagio.

OUR TOP 10

1. Arte Dolce Lyceum

1. Arte Dolce Lyceum

In one of Como's most picturesque spots – an arched facade inside the walled old town – this patisserie, ice cream parlour and delicatessen will satisfy your every craving with its freshly made, whimsical sweet and savoury offerings. Its confectionary delights are a balm for the soul: indulge in some exquisite pastries and a hot chocolate. **artedolcelyceum.it**

2. Visini

An historic delicatessen near the Duomo, recently transformed into a multifaceted experience by its second generation of owners. Come here for takeout dishes (classic Italian, vegetarian, healthy or international), artisan gourmet foods, café, wine shop, wine bar and bistrot with cookery courses. Fancy a picnic on the lake? They can prepare one for you. **visini.it**

3. Caffè Milani

With more than 80 years of expertise in importing and roasting coffee, Caffè Milani lies 4km from Como. A coffee-culture discovery experience awaits here. Browse the museum – divided into botanical 'islands' – and the technical training centre, before enjoying a freshly brewed cup of coffee, or buy some to take home. **caffemilani.it**

4. I Tigli in Theoria

Chef Franco Caffara earned Como's first Michelin star at his restaurant I Tigli al Lago in 2012, achieving the same recognition after he opened I Tigli in Theoria in 2014 at the beautifully renovated 15th-century bishop's palace. It houses an exquisite restaurant with special old-world charm, offering unforgettable contemporary cuisine. The impeccably curated multi-room interior includes one dedicated solely to aperitifs and cocktails, and there's also a delightful garden. **theoriagallery.it**

4. I Tigli in Theoria

5. Osteria del Gallo

This storied eatery in Como's centre was once a literary haunt. Today, it serves down-home cooking with a seasonal menu that uses fresh local ingredients. The warm, old-style osteria also sells wines, amaro herbal liqueurs from the Como area (try one called Piz) and regional deli products (sample some Valchiavenna bresaola or local cheeses). **osteriadelgallo-como.it**

6. Da Gigi

In the old town, Enoteca da Gigi has been passionately run by the same family since 1930. It's where the locals come to enjoy an aperitif of excellent white, red or bubbly. An array of Italian and international bottles is available to buy, or to drink at the chic wine bar or tables outside. **enotecagigi.com**

7. Le Specialità Lariane

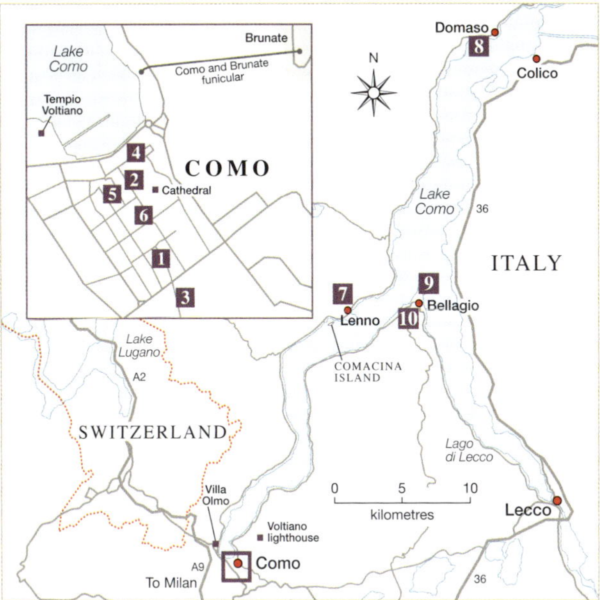

7. Le Specialità Lariane

This family business was founded by Marco Molli in 1985 in Lenno (on the lake, about a 50-minute drive from Como). It specialises in smoking and dry-curing freshwater fish, particularly the lake's iconic missoltino (shad), as well as lake-fish bottarga (roe) and marinades. You'll find all these delicacies for sale in the shop on the premises. **lespecialitalariane.it**

8. Cantine Angelinetta

Heading northwards past picturesque lakeside villages, 90 minutes from Como is Domaso. Emanuele Angelinetta's winery is perched in a striking position right above the water. Its vineyards cling to the hillside on steep terraces with drystone walls, and among them is the area's only native variety, the Verdese: be sure to taste it. **cantineangelinetta.com**

9. Silvio

Set in the gardens of Villa Melzi, 30km from Como, Silvio is a prestigious spot that has been Lake Como's premier fish restaurant for more than a century. Owner Cristian is a fisherman – he can take you on a fishing trip aboard his skiff. The house speciality is missoltini (dried shad). **bellagiosilvio.com**

10. Ristorante Mistral

10. Ristorante Mistral

With its tastefully luxurious ambience and scenic terrace overlooking the water, the Michelin-starred Mistral in Bellagio is the restaurant of one of the lake's most elegant hotels, Villa Serbelloni. Chef Ettore Bocchia uses molecular gastronomy to create innovative haute cuisine, alongside more traditional recipes. **ristorante-mistral.com**

GETTING THERE

Fly to Milano Malpensa, and you can rent a car and drive to Como in 45 minutes.

MAIN APPELLATION Terre Lariane IGT
PLANTED AREA 100ha
MAIN GRAPES Merlot, Chardonnay, Riesling, Pinot Nero, Verdese
CLIMATE Pre-alpine, with lacustrine influences

FRANCIACORTA

Most famous for its prestigious sparkling wines, Franciacorta in Lombardy is not often high on the tourist itinerary. But with its stunning hills and lake, pre-Roman history and many wineries to explore, you won't want to leave.

Monte Isola in the middle of the deep blue Iseo lake

Have dinner in any self-respecting restaurant in Italy, or a pre-dinner aperitivo in a bar there, and you'll be offered a glass of Franciacorta. The sparkling wines from this small, terroir-driven DOCG wine area in Lombardy, northern Italy, are Italy's answer to Champagne: high-quality, home-grown bubbles of prestige. If these wines are less well known outside Italy it's because the majority of the 17.5 million bottles produced in Franciacorta each year are drunk in Italy.

'Up and down the country, a glass of Franciacorta is our preferred way to celebrate or start a meal,' says Silvano Brescianini, the president of the Franciacorta consorzio. 'It's an expression of Italianità – Italianness.'

Franciacorta is as tied to Italy's national identity as Parmesan cheese or Parma ham.

There's a long history behind this loyalty: Franciacorta has been known for its wines since at least the Middle Ages, when Germanic tribe the Lombards held a seat of power in Brescia, in the southeast. The most likely origin of the region's name is from the Latin franchae curtes – 'exempt from paying taxes' – due to the tax-free zone created there in the 11th century, although some theories suggest Charlemagne named it Franciacorta, to mean Little (or short) France.

It's a scallop of land near Bergamo, less than one hour's drive east of Milan within a crescent of hills, bordered by the Lago d'Iseo lake to the north and the flat Po valley to the south.

Perfectly formed

Unlike many more sprawling wine regions, Franciacorta is compact, just 25km by 10km, with almost 3,000ha of vineyards. So it's the perfect place to spend a long weekend or take a detour for a few days from Milan. You can quickly get a feel for the landscape, visit wineries small and large and eat some great food. There's also a selection of complementary activities to make it more fun, such as horse-riding through the vineyards or exploring the **Strada del Vino Franciacorta** wine route (franciacorta.net – click on 'The road' tab) on e-bikes. The pre-alpine Iseo lake is small but spectacular, with Monte Isola, the largest inhabited lake island in Europe, at its centre. Boating is available on the lake, and those who fancy a romantic getaway can stay at lakeside hotels, a short drive from the vineyards.

Franciacorta DOCG is a sparkling wine made using the 'metodo classico' – or traditional method – during which the wine undergoes a natural second fermentation in the bottle as in Champagne (as opposed to in a large tank in the Charmat method used for Prosecco).

For Franciacorta DOCG wines, the release date cannot be less than 25 months from the harvest, and many wineries age their more prestigious wines even longer. During this long ageing process, the wines acquire complexity and staying power. As is the case with Champagne, the dosage added after disgorgement of the spent yeast deposits determines the level of dryness, ranging from extra brut to demi-sec; some are also made without dosage, completely dry. Franciacorta can be paired with a large assortment of foods, from savoury antipasti to pastas, seafoods and even some meats and cheeses.
The Franciacorta wine route weaves in and out of the vineyards, so the most direct way to visit wineries is by car. A good start is at the Berlucchi cellars (see 'My perfect day', right). Guido Berlucchi and his oenologist Franco Ziliani were the 'grandfathers' of Franciacorta, who began making sparkling wines in the area 60 years ago. Before then, still wines were the norm in the region.

The wineries making waves

'Franciacorta's character derives from its terroir – the pebbly, well-draining morainic soils that are interspersed here with marine sediments, and the lake that tempers our weather,' says Silvano Brescianini as we tour the vineyards of the Barone Pizzini estate *(baronepizzini.it)*, where he is executive vice president. 'That's why Chardonnay, Pinot Nero and Pinot Bianco do so well here.' Brescianini has also been a champion of the only local variety to be included in the blend for Franciacorta, Erbamat. This rare white grape has been known

'The sparkling wines from this small, terroir-driven DOCG wine area in northern Italy are Italy's answer to Champagne'

Osteria della Villetta

Daniele Gentile, Corte Fusia

MY PERFECT DAY IN FRANCIACORTA

Pasticceria Roberto

MORNING

Wherever I'm sleeping in Franciacorta, I have breakfast at **Pasticceria Roberto** *(pasticceriaroberto.com)* in Erbusco. The pastries and buns are excellent, including the cloud-light brioche veneziana (filled with crème pâtissière). For extra calories, try the cappuccino della nonna, enriched with egg. I'm happiest with a spremuta d'arancia, freshly squeezed orange juice. From there, it's a short drive to visit the most historic winery in Franciacorta, **Guido Berlucchi** *(berlucchi.it)*. At its heart is the handsome 17th-century palazzo where, in 1961, the first 3,000 bottles of a sparkling wine 'in the French style' were made by Franco Ziliani. Today, award-winning wines are still made by the Ziliani family and aged in the palazzo's imposing underground cellars.

LUNCH & AFTERNOON

Lunch is on the spectacular terrace of **Albereta Relais** *(albereta.it)*. Once the working home of the late great Italian chef Gualtiero Marchesi, the kitchen has maintained the maestro's focus on clean flavours and excellent ingredients and technique, even if the menu has been internationalised. If you, like me, love organic wines and heroic viticulture, the two young owners of **Corte Fusia winery** *(cortefusia.com)* focus on reclaiming abandoned hillside vineyards from which they make characterful wines, and you can arrange to walk with them in their sloping, rocky vineyards on Monte Orfano with views over the Po valley before a tasting in their courtyard headquarters. From there it's a short hop to the cellars of **1701** *(1701franciacorta.it)*. Silvia and Federico Stefini's cellars may be less picturesque, but their biodynamic viticulture and winemaking (for some of their wines) in large Italian clay jars makes this a must for natural wine lovers. You can also visit their large walled vineyard.

EVENING

I've saved room for dinner at my favourite traditional trattoria in Palazzolo sull'Oglio. Award-winning **Osteria della Villetta** *(osteriadellavilletta.it)*, which dates back to 1900, is a classic: family-run, hospitable and fairly priced. Sample wonderful home cooking and selected local wines at wooden tables in rooms that are rich in atmosphere. Just nearby I'll happily retire to **Cappuccini Resort** *(cappuccini.it)*. The former 18th-century monastery was completely abandoned until Rosalba Tonelli Pelizzari lovingly restored it – with her own artistic style – and now includes 14 rooms, terraced gardens, a restaurant and a uniquely picturesque spa.

since at least 1564 and has large, compact bunches and higher acidity than Chardonnay, so it's perfect for Franciacorta. 'Very few plants of Erbamat remained but we've been cultivating it and we now have two vineyards.' The hope is to produce even more distinctive wines from Erbamat in the future.

Pierluigi Villa, of Santa Lucia winery *(santaluciafranciacorta.it)*, is another fan of Erbamat and has played a central role in its recent history. An ampelographer by profession (one who studies and classifies grape vines), he studied local grapes in Brescia and helped to classify the variety. He even makes small quantities of a pure Erbamat sparkling wine. 'This grape's natural higher acidity means we can let it ripen longer than Chardonnay and make wines that can't be mistaken for any other part of the world.'

Impressively, 80% of Franciacorta's vineyards are now being grown organically. That includes those of the trendsetting Ca' del Bosco *(cadelbosco.com)*, where huge investment has produced a modernist cellar and sculpture park that shouldn't be missed by fans of modern art.

While the biggest estates boast showstopping cellars and landscaped gardens, it's visits to the smaller, family-run estates that are the most illuminating about the Italian way of life. Giuliana Cenci and her son Maurizio Bassi live in an 18th-century cascina, or country farmhouse, Vigneti Cenci *(vigneticenci.com)*, on the slopes of Monte Orfano. Its courtyard, with shaded tables and overhanging vine pergola, is the perfect place to taste their wines after a walk into the vineyards to see the views. 'My father started out making still wines but realised that the sparkling wines made here were more exceptional,' says Cenci. 'We're carrying on that tradition and offering the hospitality that makes Franciacorta so special.'

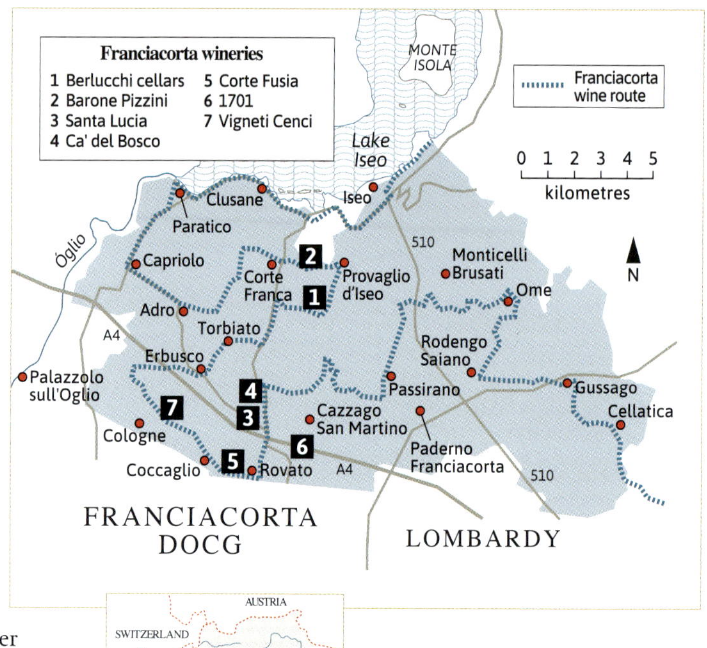

GETTING THERE

Two airports are close to Franciacorta: Milano Linate and Orio al Serio, which is close to Bergamo and operated primarily by Ryanair. From there it's easiest to rent a car.

Biking the Strada del Franciacorta wine route

Silvia Stefini, 1701

YOUR FRANCIACORTA ADDRESS BOOK

ACCOMMODATION

Agriturismo Locanda le Quattro Terre
For a restful stay immersed in the countryside at Corte Franca, this winery agriturismo offers spacious rooms, a restaurant with local dishes and easy access to the lake and wineries. **quattroterre.it**

Corte Lantieri, Capriolo
The agriturismo of a fine winery, Lantieri di Paratico, is surrounded by vineyards and has its own restaurant and pool. **cortelantieri.it**

Hotel Araba Fenice, Iseo
If it's the lake you fancy, stay at this gorgeous hotel right on the shore, with lake views and a real feeling of the Grand Tour. **arabafenicehotel.it**

RESTAURANTS

Dispensa Pani e Vini, Torbiato
Wine shop, wine bar and restaurant, this is a perfect place for a meal or for sampling wines accompanied by assorted cheeses and salumi in a handsome contemporary setting. You can also buy bottles to take away. **dispensafranciacorta.com**

Dispensa Pani e Vini

Ristorante Radicì, Iseo
In the centre of the lakeside village of Iseo, with an outdoor terrace, this is a great place to sample fresh lake fish and local pastas after a stroll along the lakefront. **radiciristoranteiseo.com**

Ristorante Dina, Gussago
A gem for fans of Italian modern cuisine: chef Alberto Gipponi's idiosyncratic five-table restaurant in a vaulted interior successfully explores textures and flavours, emotions and ideas. **dinaristorante.com**

SHOPS & MARKETS

Cantine di Franciacorta, Erbusco
This is the place to find a huge range of the area's wines at cellar prices. Great for tastings and for buying bottles to take home. **@cantinefranciacorta**

Gelateria Leon d'Oro, Iseo
On the waterfront, this is the best ice cream in the area. Don't miss their fresh fruit flavours. **@gelaterialeondoro**

Iseo market
Friday morning is the time to explore the big, busy weekly market in the streets around Piazza Garibaldi in Iseo. **comune.iseo.bs.it**

Find out more...
Details about the Strada del Franciacorta wine route, sports, hospitality and the wineries are available on the consorzio's excellent website, **franciacorta.net**

PIEDMONT

You could easily devote an entire trip to the Langhe area alone, with its endless rolling hills covered in vines, fabulous range of places to stay and world-class gastronomy seemingly at every turn. But, you can roam further afield in Piedmont to find a wealth of other great travel experiences too.

'Simply letting your gaze wander down the rows of vines provides a sense of plenitude rivalled by few places on earth'

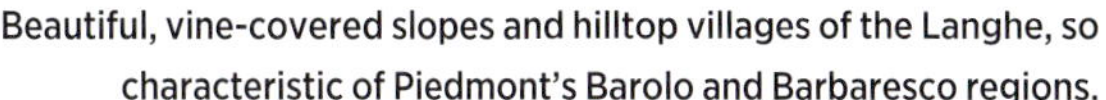

Beautiful, vine-covered slopes and hilltop villages of the Langhe, so characteristic of Piedmont's Barolo and Barbaresco regions.

Lying at the foot of the Alps in the northwestern corner of Italy, Piedmont is sheltered on three sides by mountains (the Alps and Apennines), and blessed with the ideal climate for viticulture. Here, the vines and landscapes seem to create a living painting: simply letting your gaze wander down the rows of vines provides a sense of plenitude rivalled by few places on earth.

Piedmont embodies the history and culture of Italian wine, and draws wine lovers from around the world. It is famous for its 'Three Bs' – Barolo, Barbaresco and Barbera. The first two, made only from Nebbiolo grapes, are its most prestigious wines; the third is for everyday drinking. That said, the region's sheer diversity of native varieties can leave even the most clued-up connoisseur feeling overwhelmed. The vines planted on this well-suited terroir are often centuries old, nurtured by generations of skilled wine-growers who have clung to tradition and produced some incredible wines.

Immersive Langhe

All of Piedmont's wine-growing zones make fantastic destinations for gastronomic, viticultural and cultural tours, and can be loosely grouped into four macro-areas. First, let's head into the heartland of Piedmont wine: Langhe e Roero, with its bucolic landscape dominated by hills scattered with vineyards, is on the UNESCO World Heritage List. Alongside the iconic Barolos and Barbarescos, other wines to taste here are Arneis (white), Roero (mainly Nebbiolo) and Dogliani (Dolcetto).

FACT FILE: PIEDMONT

PLANTED AREA 44,667ha
CLIMATE Continental
KEY REGIONS
Barolo DOCG, Barbaresco DOCG, Roero DOCG, Asti DOCG

KEY GRAPES
Red Nebbiolo, Barbera, Dolcetto; White Moscato Bianco, Arneis, Cortese, Erbaluce

GETTING THERE

Torino Caselle airport is located 16km outside Turin. Hire a car and head for the Langhe, about an hour to the southeast. From there you can head back up to Alto Piemonte and Novara (1.5 hours to the north), back down to Monferrato (another hour from there) and lastly to Asti (20 minutes).

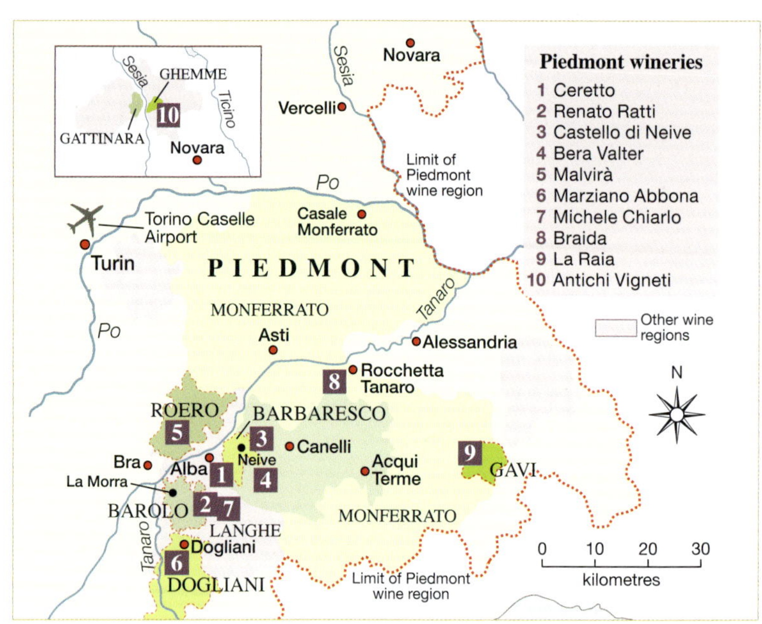

There are some 14 regional wine shops, many housed in castles and stately homes, which offer buying advice and tastings. The area is also home to the celebrated white truffle – the town of Alba holds an annual truffle festival during October and November – and there are many shops selling this local treasure. Take your love of truffles a step further and accompany a trifolau (truffle-hunter) with his faithful tabui (truffle-hunting dog), or enjoy a tasting experience: book through the Centro Nazionale Studi Tartufo *(tuber.it)*.

And be sure not to miss the Saturday food market in Alba's streets and squares. As well as truffles, one of the world's finest hazelnut varieties, the IGP Nocciola Tonda Gentile, is cultivated here and is the key ingredient for many of the local sweets and desserts.

If you love walking, the Mangialonga *(mangialonga.com)* is a 4km hike through the vineyards of La Morra, with plenty of food and wine tasting along the way. It is held each year on the last Sunday in August. Meanwhile the Collisioni festival in Barolo *(collisioni.it)* blends rock music, literature, food and wine. Major performers from around the world take part in it every July.

Basilica di San Gaudenzio

What else to do in this part of Piedmont? You can wander among the castles, through the villages and museums (particularly the WiMu wine museum and Corkscrew Museum, both in Barolo), and marvel at the panoramic views over the Langhe.

So much to see

Venturing northwards, Alto Piemonte Nebbiolo creates fresher, less structured wines – particularly in Gattinara and Ghemme – than its cousins in the Langhe. You'll also find white wines made from the Erbaluce grape, in sparkling, dry still and dessert passito styles. You could visit Novara, where the Basilica di San Gaudenzio *(turismonovara.it)* is famous for its monumental 121m-high cupola designed by Alessandro Antonelli and added in the 1880s – climb up for sweeping vistas over the city as far as Monte Rosa in the Alps to the northwest; while the city's Castello Visconteo Sforzesco *(ilcastellodinovara.it)*, is set in one of Piedmont's loveliest parks.

Heading back south, the next stop is in the third macro-area, Monferrato. Another World Heritage Site, famous for its Infernòt (series of unique, hand-dug wine cellars), this is the

A historic medieval parade during the Alba Truffle festival

home of Barbera, Dolcetto and Cortese, but also the birthplace of Italy's earliest sparkling wines in the last of our macro-areas: Asti. As you explore the vast underground cathedrals dug into the tufa in Canelli, and meander around castles, craft shops and cultural sights, make sure to enjoy some Moscato d'Asti, Barbera d'Asti and Ruchè di Castagnole.

WINERIES TO VISIT

Antichi Vigneti di Cantalupo

Ghemme is at the heart of northern Piedmont's time-honoured wine country. Here, Alberto Alunno has managed Cantalupo since 1981 and is devoted to his native terroir, which boasts an exceptional – albeit not unique – mineral complexity that imbues Ghemme Nebbiolos with finesse and elegance. Alunno's wines are a prime example – treat yourself to a taste of one of the older vintages. The estate's nerve-centre is its spectacular 1,200m^2 cellar built beneath the hillside: an underground amphitheatre of wide steps stacked with barrels of various sizes mirrors the gradient of the slope above. Upon entering, you are met with an awe-inspiring view of all the precious casks displayed in this 'theatre of ageing'. A door leads to a corridor lined with small cells that can each store up to 5,000 bottles. Divided by cru, the bottles slumber there for at least a year, nestled in red velvet. *cantalupo.net*

Valter and Alida Bera with their sons Umberto and Riccardo

Bera Valter

If Neviglie has become a favourite port of call for many Moscato lovers, it owes much of its fame to the efforts of the Bera family, which stretches back centuries. The Bera winery sits on the road leading to the medieval village, immortalised by the Piedmontese writer Beppe Fenoglio. It's worth visiting for a stroll around the winding streets and its sweeping views of the surrounding hills. The manor house stands majestically on the vine-clad hill to the left of the road, with the new house and modern, recently expanded winery below. If you're after a superb Moscato d'Asti, look no further – Bera is a standard-bearer. Valter began bottling it in the 1970s, and now works with his sons Umberto and Riccardo. While best known for their Moscato, more recently the family has branched out to make Nebbiolo-based wine too. *bera.it*

Braida

As a local singer-songwriter put it, 'this town has no surprises: a church, six houses and 10 vineyards' – but there are plenty of reasons to visit Rocchetta Tanaro. Here, Giacomo Bologna 'Braida' (he lent his family nickname to the winery) made history with his innovative cask-aged Barbera. You'll be able to taste a range of the estate's Barberas, from the lightest, most frivolous offerings to more lush and complex ones. With his irresistible friendliness, enthusiasm and empathy, Giacomo Bologna put Rocchetta Tanaro on the map; a town where one can still sense the infectious personality of this special man who adored his land, its wine (made from Barbera), his family and friends. His two children, Raffaella and Giuseppe, are both oenologists and have inherited their father's values, smilingly describing themselves as 'dynamic conservatives'. The warm welcome you'll be given by this remarkable family is as moving and memorable as the wines themselves. *braida.it*

ABOVE Giuseppe and Raffaella Bologna, brother and sister owners of Braida

Castello di Neive

Set among rolling, vine-covered hillsides is the 18th century Castello di Neive, with dreamy views across the landscape. Close to Barbaresco, Neive is one of Italy's prettiest villages. In the 19th century, oenologist Louis Oudart (later employed by the Savoy royals to help create the first Barolo) created some fine red wines in the castle cellars. The Castello's more recent history began when the Stupino family bought the property in the 1960s. Today, the affable Italo runs the estate – he helped bring about the comeback of the white Arneis grape and began a clonal selection programme here in 1982. This winery is worth visiting just for the stunningly beautiful castle itself, its vaulted cellar filled with wooden casks. But taste its Barbaresco, from the monopole cru of Santo Stefano – one of the most important vineyards in the Langhe. *castellodineive.it*

Ceretto

From its headquarters at Tenuta Monsordo Bernardina, Alba, the Ceretto family oversees three other wineries (in Castiglione Falletto, Barbaresco and Santo Stefano Belbo), covering a total of 160ha, with vineyards in the most prestigious crus. Patrons of the arts, the Cerettos are also passionate foodies, operating two restaurants and a patisserie, where the star ingredient is Piedmont hazelnuts.

Visit for not one, but several unique experiences. Take in the giant transparent 'grape' jutting out over the vineyards at Alba, for wine tastings immersed in nature. Equally striking is the clear glass cube set into the hilltop at Bricco Rocche; and the Chapel of Barolo at the Brunate vineyard, reinterpreted by artists Sol LeWitt and David Tremlett. Alessandro Ceretto, the

LEFT Italo Stupino, who co-owns Castello di Neive with siblings Anna, Giulio and Piera

Ceretto's transparent 'grape' tasting suite juts out over its vineyards at Alba

third generation of the family at Ceretto, began converting the vineyards to organic farming in 2010. They received certification in 2016, and all the crus are now farmed biodynamically. The wines are pleasingly stylish, elegant and enjoyable. *ceretto.it*

La Raia

A microcosm of wellbeing and an oasis of biodiversity: this is La Raia. Owned by the Rossi Cairo family, this enchanting place allows visitors to revel in nature as they sample a selection of impressive Gavi wines. Step inside and the experience will stay with you forever. The project began in 2003 in the heart of Gavi, home of the Cortese grape. The estate now includes a Demeter-certified biodynamic winery; a farmstay with restaurant, spa and pool; a reimagined Italian garden; a park with display of contemporary artworks, and a cultural foundation. Its 180ha comprise 45ha of vineyards, along with arable land, cattle pastures, hazelnut groves and woods of chestnut, acacia, elder and oak, which provide habitats for numerous wildlife species. The winery's one-of-a-kind cellar features a glass wall and rammed earth construction, built using the age-old, eco-sustainable pisé technique, which lends its name to the estate's Gavi cru, Pisé. *la-raia.it*

'revel in nature' at La Raia's eco-friendly farmstay restaurant

Malvirà

This family-run estate was founded in 1974 on just 2ha. Today it covers 42ha. In Piedmont dialect, Malvirà means 'badly turned' – relating to the estate's original plot which, unlike its new location, was north- rather than south-facing. The estate is one of the leading producers of Roero, with plots in some of the most renowned crus, from Mombeltramo to Renesio, Saglietto, San Michele and Trinità in the Canale area. The second-generation Damonte brothers – Roberto looks after the vineyards and Massimo the cellar – are now flanked by their sons, Giacomo and Francesco. With their unwavering passion for promoting their terroir and pursuing quality, the Damontes are an inspiration to all local growers. Certified organic in 2014, the estate is situated at the foot of the Trinità vineyard, and sprawls around the stunning Villa Tiboldi. Here, the family welcomes wine lovers to its guesthouse, complete with restaurant and pool. *malvira.com; villatiboldi.com*

Marziano Abbona

Celso Abbona was one of the first in his generation to believe in the grape that thrives on the hills of Dogliani: Dolcetto. Hardly surprising, then, that his son Marziano, the owner and founder of this winery, named its flagship wine Papà Celso. With more than 50 harvests under his belt, Marziano is a larger-than-life character on Piedmont's wine scene – a chat with him, complete with anecdotes and facts, offers crucial insights into the heritage of this unique zone. The winery has become a beacon of winemaking and hospitality, with its underground cellar surrounded by a ring of hills overlooking a lake, an attractive brick-vaulted tasting room and the new farmstay. Dogliani, with its hilltop castle towering above the old town, is worth visiting, as is the Bottega del Vino Dogliani, where you can sample Dolcettos from about 45 local producers. *abbona.com*

Major art installations at Michele Chiarlo's Art Park La Court

'Art Park La Court is the largest open-air museum in a vineyard, with works by world-renowned artists and sculptors along a magical, immersive art walk'

Michele Chiarlo

Few names in winemaking have forged the history of an entire terroir quite like Chiarlo. With roots firmly planted in the Asti hills, this family has always been passionate about Barbera, but over the years it has also cherry-picked the finest crus in Langhe, Monferrato and Gavi. The must-see winery has a vertical lawn on its façade, which hints at the family's interest in preserving the equilibrium of nature. The property often hosts art shows, and 10 minutes away is an awe-inspiring place: the Art Park La Court is the largest open-air museum in a vineyard, with works by world-renowned artists and sculptors arranged along a magical, immersive art walk set against magnificent landscapes. In Cerequio, Palás Cerequio (50 minutes from the winery) is the first resort hotel dedicated to Barolo crus: its Chiarlo cellar is a treasure trove. *michelchiarlo.it*

Renato Ratti

One of the forefathers of Barolo, Renato Ratti was the first to craft a single-vineyard Barolo; to draw the map of historic vineyards; to invent the Albeisa bottle. And as president of the Barolo consorzio, he helped draft the DOCG bylaws. He began making wine in the early 1960s at L'Annunziata, a Benedictine abbey. The complex houses the Museo Ratti, with displays of ancient viticultural and vinification tools.

Ratti's son Pietro had the winery redesigned, and his love and respect for the region shines through in the way the winery blends into the landscape, its flowing lines echoing the rolling hills. Admire the Conca and Marcenasco plots from the tasting room's floor-to-ceiling windows as you sample wines that bear the true hallmarks of this terroir. *renatoratti.com*

BELOW FROM LEFT Dogliani producer Marziano Abbona with (from left) wife Bruna, daughters Mara and Chiara

The winery at Renato Ratti is designed to blend into the landscape

YOUR PIEDMONT ADDRESS BOOK

Langhe Country House

ACCOMMODATION

Bogogno Golf Resort

In Bogogno commune, in the northern part of the region, this eco-sustainable, low-impact resort is ideal for sports and exercise (golf, tennis, five-a-side football, a gym and a pool), relaxation in the wellness area and spa, and fine-dining while admiring spectacular views of Monte Rosa. Rooms are large and comfortably appointed. **bogognogolfresort.com**

Langhe Country House B&B

Set in beautiful gardens, this boutique hotel in Neive has been expertly renovated in rustic-chic style. Its six suites blend modern amenities and old-world charm, topped off by Nadia and Alessandro's superb hospitality. Unwind by the pool, or try your hand at a cookery course. **langhecountryhouse.it**

Locanda del Pilone

An exquisitely restored farmhouse outside Alba, warm and elegant in equal measure. Bright, spacious rooms overlooking the vineyards feature antique furniture. Take a dip in the pool and dine in the Michelin-starred restaurant. **locandadelpilone.com**

La Piola

WHERE TO EAT: BREAKFAST

Pasticceria Barbero

Its windows look out from beneath the porticoes in the centre of Cherasco. The house speciality here is the legendary Baci di Cherasco hazelnut-chocolate pralines, invented by Marco Barbero, who founded the patisserie in 1881. **barberocioccolato.it**

LUNCH

La Piola

La Piola is a traditional, friendly osteria serving local fare. Don't be fooled by the chalkboard menu: this bright, appealing eatery in Alba uses the same suppliers as three-star Michelin Piazza Duomo upstairs, and chef Enrico Crippa is behind both projects. Order the classics: the Piedmontese antipasto selection and trolley of boiled meats with sauces. **lapiola-alba.it**

Repubblica di Perno

A cosy, authentic osteria in Monforte d'Alba offering top-notch Langhe cuisine. Order the agnolotti del plin (stuffed pasta), seasonal vegetables with bagna cauda (a hot garlic and anchovy dip) and the finanziera (chicken and beef sweetbreads). Reservation only. **repubblicadiperno.it**

DINNER

La Ciau del Tornavento

With vineyard views and a cellar of some 65,000 bottles, this Michelin-starred restaurant in Treiso is a gourmet mecca thanks to its chef- owner Maurilio Garola. French influenced Piedmontese classics include truffle and fish dishes. The tasting menu is pure delight! **laciaudeltornavento.it**

La Madernassa

Michelangelo Mammoliti is one of Italy's most talented young chefs. This two-star Michelin restaurant in Guarene offers creative cuisine, flavoured with herbs from the resort's own kitchen garden. **lamadernassa.it**

SOUTHERN ITALY

Italy's southern regions have varied cultures but are united by their passion for wine and hospitality. From seaside villas with vineyards to winery-owned hotels amid ancient caves, Carla Capalbo shares her top wine destinations in the south.

The pool at Capofaro Locanda on the island of Salina

The medieval village of Civitella d'Agliano, home of La Tana dell'Istrice

Any lover of Italian wine who also loves to travel will have undoubtedly visited Tuscany and, hopefully, Piedmont in their search for wonderful places to stay on wine estates. Far fewer have explored the fantastic regions of Italy's south, below Rome. I'm passionate about these southern regions. To me, they express the most quintessentially Mediterranean aspects of Italian culture – not only for their sun and sea, but also for the rich layers of culture that have been left there by thousands of years of occupation, from the Greeks and Byzantines to the Arabs and Bourbons. Pick any one in my selection of fabulous places to visit, each with a link to wine, and you will leave seduced by the food, wine and hospitality of the Italian meridione.

Sergio Mottura, La Tana dell'Istrice

CIVITELLA D'AGLIANO, LAZIO

Sergio Mottura's winery is about 90 minutes' drive south of Rome, at Civitella d'Agliano, in the beautiful post-volcanic landscape that characterises so much of central Italy. The estate's headquarters are in a handsome villa in the heart of the medieval village, a short distance from its organic vineyards featuring the white Grechetto and red Montepulciano d'Abruzzo varieties, among others.

Named La Tana dell'Istrice ('the porcupine's lair'), the family's spacious villa has been converted into 11 rooms for guests, without losing sight of its historical origins. The pretty dining room and well-equipped kitchens offer lunches and dinners by appointment. Children are welcome too, and will find the large swimming pool set in the midst of the vineyards irresistible. There are lots of optional activities, from wine tastings and cooking classes to day trips, as well as the chance to experience the grape and olive harvests in season. Best of all is the proximity of the Mottura family: Sergio and his sons are gracious hosts and bring this slice of la dolce vita to life.

ergiomottura.com

Feudi di San Gregorio

SORBO SERPICO, CAMPANIA

Feudi di San Gregorio winery has long been a beacon of stylish modernity in the rural hills of the Campanian hinterland. Less than an hour's drive east from Naples, the landscape changes as it begins to climb towards the upper reaches of the Apennines, the Italian peninsula's 'backbone'. Set on a high point above Sorbo Serpico, a few kilometres from Avellino, with stunning views of hills and vines, the winery's central buildings were designed by the Japanese architect Hikaru Mori in 2001. She brought a pared-down, elegant aesthetic to an area best known for its rusticity. The graphic design of the late Massimo Vignelli complemented her minimalist directive and gave Feudi its unmistakable look.

Marennà restaurant at Feudi di San Gregorio

A room at Il Palazzotto Residence & Winery in Matera

Guest houses at I Cacciagalli in Campania

Visitors can tour the cellars and vineyards, view its modern art installations, learn about the low-impact approach the winery now practises, and eat in the award-winning panoramic restaurant, Marennà. Here the food features Campanian ingredients, Neapolitan traditions and dishes that complement the estate's wines, from the three classic local DOCGs – Fiano di Avellino, Greco di Tufo and Taurasi – as well as from more recent projects, like the sparkling Dubl wines from native grapes vinified in the style of Champagne. ***feudi.it***

Il Palazzotto Residence & Winery

MATERA, BASILICATA

This extraordinary hotel in the ancient cave city of Matera is owned by the Francesco Radino winery. The winery's estate and vineyards are located at Rionero in Vulture, about 90 minutes' drive from Matera, where the D'Angelo family – who bought the winery in 2015 – produce organic wines from Aglianico and other local grapes.

The Sassi, as the city's cave dwellings are called, run down through a canyon and were inhabited continuously for centuries – if not millennia – until the 1950s, when the inhabitants were moved out due to abject poverty. After careful restoration, Matera was made a UNESCO World Heritage Site in 1993.

The city has now been brought fully back to life, and this hotel is an example of beautiful design that enhances but does not overpower the ancient structures.

Taste the family's wines in a spectacular underground wine lounge complete with limestone carvings and arches. If you feel like splurging, opt for one of the suites, as they occupy the most stunning spaces. The hotel is within walking distance of the cathedral, and the city's lively central streets with their many restaurants and shops. Matera was a joint European Capital of Culture in 2019, and repays any visit with an unforgettable experience. ***ilpalazzottomatera.com***

I Cacciagalli

TEANO, CAMPANIA

For lovers of natural wines, this biodynamic estate in the province of Caserta (northwest of Naples) offers a stylish yet affordable place to stay with the family. The look is spare but well designed, with wrought iron, wood and pale natural fabrics setting the tone. The pool has been landscaped to look more like a small lake, and the house accommodations are set in pretty countryside.

The wines are made in large clay amphorae by Mario Basco, and he and his young family live on the property and look after the guests themselves. They grow the local varieties of this post-volcanic area, including Aglianico, Falanghina, Fiano and Piedirosso. In the restaurant, ingredients are sourced from local organic producers and meals are served in an attractive dining room.

This is a wonderful part of the country to explore, with the majestic Reggio di Caserta – a royal palace designed by Vanvitelli for the House of Bourbon and based on Versailles – not far away. ***icacciagalli.it***

Vinilia Wine Resort

Vinilia Wine Resort

MANDURIA, PUGLIA

If Primitivo is your favourite grape, Manduria is a great place to find it. The sun-baked flat vineyards, often with bush vines stretching right down to the sea, have an ancient appeal to them: a testament to their Magna Grecian heritage. The landscape here is punctuated by Baroque churches, stone trulli, centennial olive trees and defensive watchtowers once used for sighting Ottoman and Saracen marauders. Manduria is 35km from Taranto and 50km from Brindisi, on Italy's Puglian 'heel', and makes an excellent base from which to explore both coasts.

Vinilia Wine Resort is located here, in an imposing, early 20th-century stone castle. The handsome villa has been converted into a comfortable hotel and spa with its own Michelin-starred restaurant, Casamatta, that features modern Puglian cooking. There's also a large pool for relaxing on hot days.

While the resort's vineyards are situated a few kilometres away, the town of Manduria is well worth visiting and has an interesting wine museum dedicated to the culture of its native grape, Primitivo. There are fabulous beaches nearby, as well as villages and local wineries to explore. ***viniliaresort.com***

Capofaro Locanda & Malvasia

SALINA, SICILY

The Tasca d'Almerita family has long been considered the royalty of Sicilian winemaking. Its headquarters are in the Sicilian heartlands at Regaleali, but in recent years its estates have expanded into other parts of Sicily. The jewel in that crown is Capofaro on the island of Salina, one of the volcanic Aeolian islands that belong to Sicily.

Capofaro is the perfect idyllic getaway for wine lovers. The 27 rooms, each with its own entrance, are built among vineyards where the grapes for the delicious dessert wine, Malvasia delle Lipari, are grown. The estate overlooks the sea, so there are beaches nearby, plus a central pool at the resort itself. The restaurant offers the best of the Mediterranean: fresh seafood, sun-nourished vegetables and the accents – like capers, olives,

Capofaro Locanda

anchovies and wild herbs – that give Sicilian food its distinct character. The chef, Ludovico De Vivo, creates his recipes from the many cultural influences that form Sicily's well-flavoured cuisine, including rustic peasant dishes and aristocratic food from the region's golden age. For those who want to learn how to make them, cooking classes are available on demand. Day trips to the other islands are also available, as are tours of Salina, and yoga retreats. ***capofaro.it***

Planeta, La Foresteria

MENFI, SICILY

Planeta was the first winery in Sicily with a vision to communicate the island's viticultural greatness to a modern international audience. The Planeta family has always understood the value of Sicily's diversity and has been enthusiastic in helping to build wine tourism on the island through its hospitality.

The winery headquarters are in Menfi, on the southwest coast of Sicily, and that's where the Planetas have created their country house hotel (they also have seven rooms in central Palermo). La Foresteria offers 14 rooms, a stunning infinity pool, scented herb gardens and beach access.

A relaxed, country-chic aesthetic runs through the bedrooms, the large kitchen and reception rooms. There's great food to be had, with cooking classes on offer – as well as wine tastings from all of the family's estates. In warm weather, eat outside on the terrace overlooking the vineyards. Day trips include the Greek temples of Selinunte and Segesta, the olive groves of Belice, the fish market of Mazara del Vallo and explorations of the cultural centre and salt flats of Marsala. Planeta can also provide wine tours to its other estates at Noto and on Mount Etna. ***planetaestate.it***

Argiolas

SERDIANA, SARDINIA

The Argiolas family has been the leading light in Sardinian wine for three generations. It helped the world discover native grape varieties such as Vermentino and Cannonau, and has consistently won awards for its wines.

Recently, the family has enlarged its hospitality portfolio, and now offers the chance to visit the winery and vineyards... by Segway, if you dare! You can even have an aperitivo in the vineyard, to enjoy with local cheeses and salumi. There's also an experimental vineyard of unusual native grapes on show and, in season, the chance to see the verdant olive groves.

For those more interested in food, the estate's restaurant serves Sardinian specialities, with the possibility of getting a cooking lesson from the chef. ***visitargiolas.it***

VENICE

A glittering maze of canals framed by decorative Venetian Gothic buildings, the 'floating city' of Venice fascinates the world with its water – but what of the wine?

Venice is bliss for wine lovers. Driving is out of the question for a start, and the day is traditionally punctuated by breaks at the city's many bacari (wine bars) for a small glass with cicchetti, the bite-sized snacks that Venetians do so well. The wine choice is largely local, popular varieties including the fruity Manzoni Bianco, dry Tai (also known as Friulano or Sauvignon Vert) and juicy red Raboso. The role of the landlord, or oste, is even more central here than elsewhere – and be they brusque, friendly or eccentric, they're sure to leave an impression.

Together with its outlying islands, Venice has a deeply ingrained winemaking history, and grapes continue to grow in gardens and courtyards. You can even eat under vines at restaurants such as Corte Sconta (see No3, below) and Pizzeria alla Strega *(@alla.strega.venezia)*, where a rare Bacò vine provides a courtyard canopy.

Twelve years ago, the Consorzio Vini Venezia launched a project to safeguard the city's forgotten vines, unearthing an incredible 70 varieties, 18 of which now grow in a vineyard behind the 17th-century Carmelitani Scalzi church. The vines are part of the walled Giardino Mistico *(giardinomistico.it)*, laid out in seven areas representing the order's teachings and also home to olives, woodland and herbs such as Melissa moldavica, used for a herbal tonic made here since 1710. The vines include six types of Malvasia, once so important locally that the most prestigious wine shops of 16th-century Venice were known as malvasie, specialising in wines from Greece.

Malvasia Istriana, together with Glera, now grows at another vineyard, the city's oldest, bequeathed in 1253 to Franciscans who built their San Francesco della Vigna church around it. Since 2019, the vineyard has been owned by the Santa Margherita group (famed for its Prosecco and Pinot Grigio), which is also restoring the original chapel.

No stay in Venice is complete without a trip to the islands, especially the trio where the lagoon's first, 5th-century settlers lived: Torcello, home to the famous Locanda Cipriani and a Byzantine basilica with glorious mosaics; Burano with its brightly painted houses; and peaceful Mazzorbo linked by bridge to Burano.

It's here on Mazzorbo that the Bisol family runs the lovely Venissa wine resort (see p95). Surrounded by sea, with a clever drainage system to protect the vines, the resort's walled vineyard of traditional but rare Dorona vines produces two intense, deep-golden wines made with extended skin contact, including Venissa, sold in 50cl bottles with gold-leaf labels (2014, £150 Fine & Rare).

Astonishing at any time of year, Venice's charm is only enhanced by winter mists when the welcoming glow of cosy interiors becomes magical, even more so during carnival (next one, 22nd Feb - 4th March 2025), when curiously costumed figures are a regular sight. Wander the city, explore the quiet side-canals and embrace the voluptuous Venetian hospitality when you come upon another wonderful bacaro.

OUR TOP 10

1. Cantina Do Mori

Step back in time at Venice's oldest bacaro, dating from 1462. Order a local white, such as Manzoni Bianco, Verdiso or Tai, at the long wooden bar and perch on a stool to ponder which of the array of cicchetti bites to try next: perhaps the typical creamy baccalà mantecato cod with polenta, the baby octopus or salami crostini.
Sestiere San Polo, 429

2. Cantinone Gia Schiavi

One of Venice's best-loved historical bacari is also a well-stocked wine shop specialising in bottles from northeastern Italy. Join the locals for a glass and selection of delicious cicchetti such as octopus and celery, cuttlefish and samphire or the traditional sarde in saor (marinated sardines). Standing room only, but you can perch out on the canalside, not far from the San Trovaso gondola boatyard. **cantinaschiavi.com**

1. Cantina Do Mori

3. Corte Sconta

A menu of top-quality seasonal seafood such as fried soft-shell crabs, monkfish, tuna or bream, and delicious homemade bread and pasta, in a charming location with a simple interior and a wonderful 150-year-old Glera vine providing a canopy for the internal courtyard garden. The wine list presents an ever-changing selection of labels from small-scale ethical producers selected by owner Marco Proietto. **See Facebook**

4. La Sete

Just off the bar-lined Fondamenta Misericordia and behind the same management's Da Rioba restaurant, this cosy, recently opened brick and wood-beamed bar specialises in artisan wines from all over the world – about 10 are available by the glass at any time, always including something local. Owner Tommaso Milner's family land on Sant'Erasmo island also provides fresh veg for the dozen other restaurants of the island's Osti in Orto horticultural project. **lasetevenezia.com**

5. Mercato Di Rialto

Take the Santa Sofia traghetto (gondola ferry) from the Cannaregio district to this historic market. This is where restaurant chefs source the freshest fish and vegetables, including the celebrated artichokes from Sant'Erasmo, the island known as Venice's vegetable garden. While here, shop for regional cheeses such as Morlacco and Asiago at Casa del Parmigiano and stop for wine with mini panini next door at tiny Al Mercà. **Campo Bella Vienna, 213**

6. Osteria I Rusteghi

Take a seat in the cosy interior of this upmarket version of the typical bacaro and order choice charcuterie from small producers, Prosecco-infused or grape skin-aged cheeses and something from the small daily menu, which might include scampi in vermouth or pasta with mullet roe. Ask owner Giovanni about the extensive wine collection and round off with buranelli biscuits dipped in a luscious local passito. **airusteghi.com**

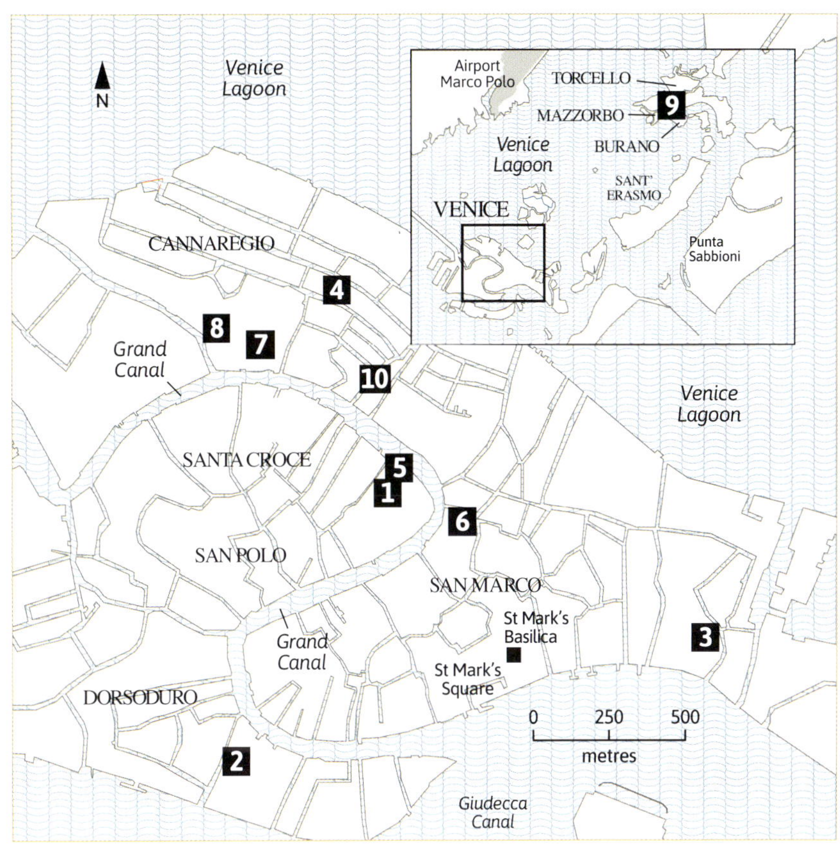

7. Zizzo

Welcoming aromas hit you as soon as you enter the San Leonardo branch of this family-run bakery, as it's here that the baking takes place for many of Venice's restaurants and hotels. Try seasonal favourites such as focaccia Veneziana at Christmas, melt-in-the-mouth frittelle (fried dough balls with dried fruit) at carnival time, and simple buranelli and zaletti biscuits all year round, ideal with Verduzzo or Torcolato passito wines. **rizzovenezia.it**

8. Trattoria Alla Palazzina

An atmospheric, wood-panelled space featuring a piano and a double bass, occasionally played by talented diners, and a spacious patio garden shaded by a leafy pergola make this a welcome refuge from the busy route through the Cannaregio neighbourhood. Relax over a mixed seasonal seafood starter, perhaps cuttlefish in black ink sauce or bigoli in salsa (thick spaghetti with sardines and onion), or simply stop for a drink during the afternoon. **ristoranteallapalazzina.it**

9. Venissa

The relaxing atmosphere on the island of Mazzorbo is even more all-enveloping here. Rare Dorona vines grow in a historic walled vineyard that's open to all for peaceful strolls and shares the space with pensioners' allotments, the chef's garden and a landmark bell tower. Eat at the one-star Michelin restaurant or informal Osteria, and book to stay at one of its five charming rooms overlooking the lagoon or the vineyard. **venissa.it**

10. Vini Da Gigio

Paulo Lazzari's thousand-strong wine collection features about 40 available by the glass, and his sister Laura serves tasty seasonal specialities from the open kitchen. Favourites include crab risotto, fresh artichokes and asparagus from Sant'Erasmo island, duck and smoked spaghetti carbonara with tuna. A truly warm, friendly welcome, with centuries-old columns and beams sets the scene for memorable meals. **vinidagigio.com**

Vines within San Francesco della Vigna church

SPAIN & PORTUGAL

Rioja's modern architecture and historic wine estates are an enduring draw for visitors to Spain. But there are also delights awaiting wine lovers in the bustling capital city Madrid, and the seaside resort of Cádiz, in Andalucia. In southern Portugal, meanwhile, the food and wine scene in the Algarve is better than ever, offering another reason to visit the popular beach holiday destination.

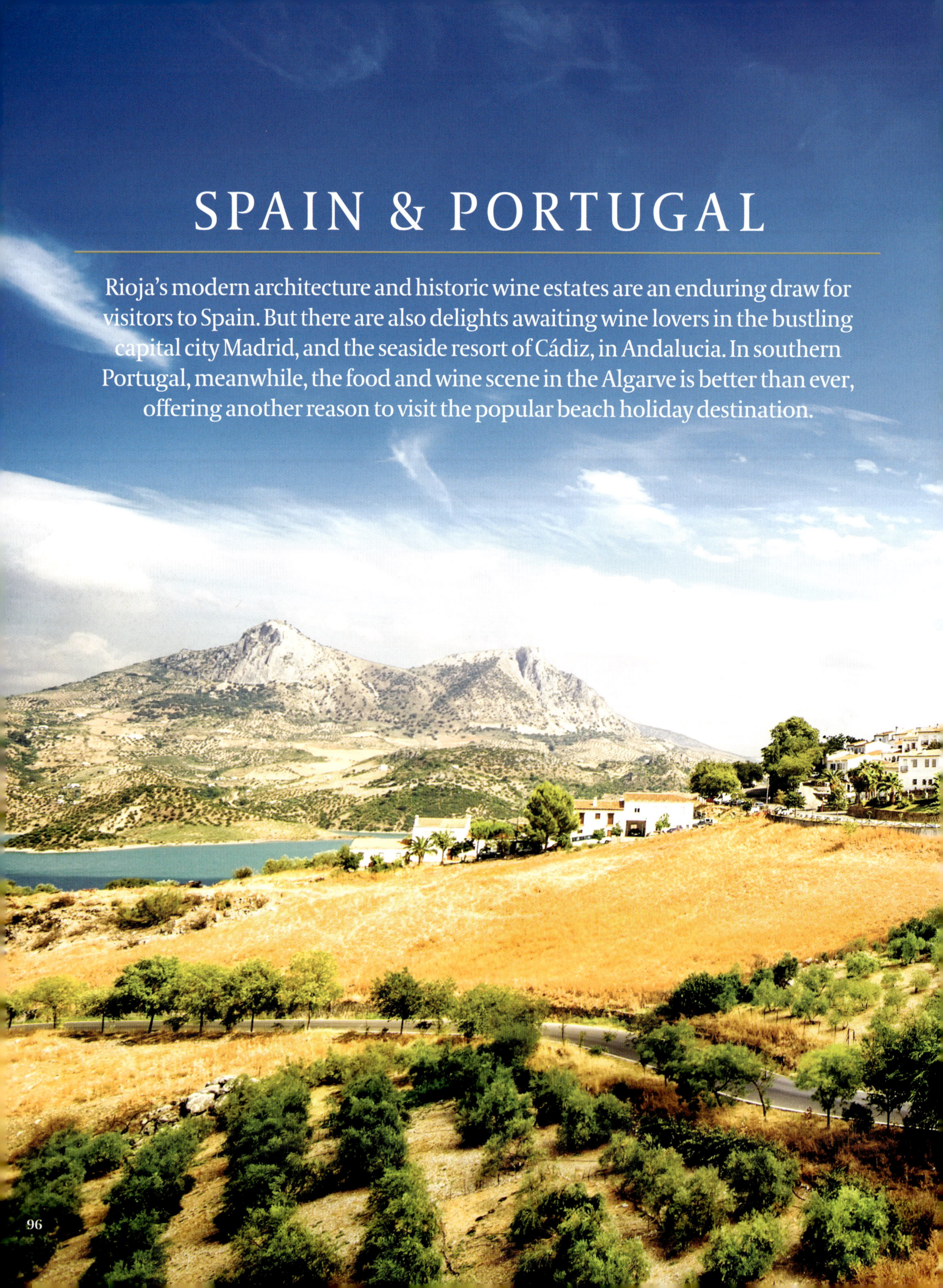

CÁDIZ

A frequent movie stand-in for Cuba and well known for its annual winter Carnaval, the city of Cádiz is also a perfect base for exploring Andalucía's fascinating wineries and tasting the region's fine food.

The rooftops and old merchants' watchtowers of Cádiz city, with a view of its cathedral and sandy beaches beyond

Looking out across the rolling immensity of the Atlantic ocean, the city of Cádiz, perched on the southwest coast of Spain, is one of the oldest inhabited cities in Europe. Situated at the end of a long sand spit on what was once an island, it has been like a little world unto itself since the Phoenicians first came here some 3,000 years ago.

While Cádiz offers fewer major monuments than some of its counterparts, there is still history and architecture aplenty. From the cathedral and the recently uncovered Roman theatre in the old Barrio del Pópulo, to the famed fish market and merchants' watchtowers, the city has a character all of its own. Stroll along the seaside garden promenades and through the gracious squares dating from the city's 18th-century economic heyday to the iconic Balneario (bathhouse) on Caleta beach, with its tiny fishing boats anchored in the bay, and the stone causeway leading to the 1706-built Castillo de San Sebastián fortress.

To the south of the city, heading out into the province of Cádiz, the white sand beaches of the Costa de la Luz stretch as far as Tarifa at the southern tip of Spain, not far from Gibraltar. In late

spring and early summer this stretch of coast is where the annual tuna catch, the Almadraba, takes place, and a visit to the local fish markets in Conil (about 43km down the coast from the city by road), Barbate (66km) and Zahara de los Atunes (75km) is well worth the effort, as are the Roman ruins of Baelo Claudia in Bolonia, between Zahara and Tarifa. In the nearby Sierra de Grazalema mountain range you'll find a number of Andalucía's famous white villages perched on rocky hilltops, including Vejer nearer the coast (62km), and inland Arcos de la Frontera (65km), while in the north of the range sit Grazalema itself (111km) and Zahara de la Sierra (116km).

Wine country

In the northern part of the province, bordered by the towns of Jerez de la Frontera, El Puerto de Santa María and Sanlúcar de Barrameda, is the famous 'Sherry triangle' with the production of all Sherry wines governed by the Denominaciones de Origen of Jerez and Manzanilla. The Vino de la Tierra de Cádiz regional appellation oversees the production of unfortified still wines (vinos de pasto) throughout the province, and is gaining a reputation for the quality and individuality of its wines, with some bodegas using native grape varieties such as Tintilla de Rota and Uva Rey.

Just outside Jerez is Bodegas Luis Pérez (*bodegasluisperez.com*), founded in 2002 by Luis Pérez Rodriguez and his family with the aim of recovering some of the indigenous grape varieties of the region to produce both Sherry and non-fortified wines. The original Vistahermosa farmhouse, located in the countryside and surrounded by its own vineyards on the El Corchuelo estate, is still in use, though a new modern winemaking and office facility has been built alongside. Visits include a tour of

'From the cathedral and Roman theatre to the fish market and merchants' watchtowers, the city has a character all of its own'

LEFT Sherry sampling at Bodega Gutiérrez Colosía
BELOW The hilltop town of Zahara de la Sierra, in the Sierra de Grazalema mountain range

MY PERFECT DAY IN CÁDIZ

MORNING

I always love starting the day with a market visit, and the Mercado Central in Cádiz is one of my favourites, but today we have to be at the Maritime Port Terminal *(cmtbc.es/catamaran.php)* to catch the 9.55am catamaran to El Puerto de Santa María. A pleasant 35-minute glide across Cádiz bay, and then just up the Guadalete river, leaves you pretty much at the doorstep of **Bodega Gutiérrez Colosía** *(gutierrezcolosia.com)*. Founded in 1838, this classic-style bodega, with its arches and high ceilings (affectionately known as Sherry cathedrals) has been owned by the Gutiérrez family for more than 100 years. Visits are available Monday to Saturday at 11am (booking required) and include a winery tour and Sherry tasting.

LUNCH & AFTERNOON

After your bodega visit you'll have time for a quick Sherry cocktail next door at **Bespoke** *(bespokepuerto.com)*, run by daughter Carmen Gutiérrez, before heading to local seafood and wine mecca **Restaurante El Faro de El Puerto** *(elfarodelpuerto.com)* for an unforgettable lunch. Chef and owner Fernando Córdoba not only offers one of the best traditional menus in the province of Cádiz but, since opening in 1988, he has amassed an incredible wine cellar of more than 800 references, mostly Spanish, with about 220 available by the glass. Sommelier Javier Manso is on hand to guide you; when you book, you can request the option of starting your meal with aperitifs inside the magnificent 150m^2 bodega.

A starter at El Faro de El Puerto

EVENING

After lunch, a short train ride will bring you back to Cádiz. Once rested up, start off your evening with a glass of Sherry straight from the barrel at **Taberna La Manzanilla** on Calle Feduchy before stopping in at **Mesón Cumbres Mayores** *(cumbresmayores.com)* for its exquisite Ibérico pork dishes. One of the many things I love about Cádiz is that nothing is very far to walk and that any route you take back to your hotel will include pretty streets, parks, squares and perhaps a seaside stroll. Perfect.

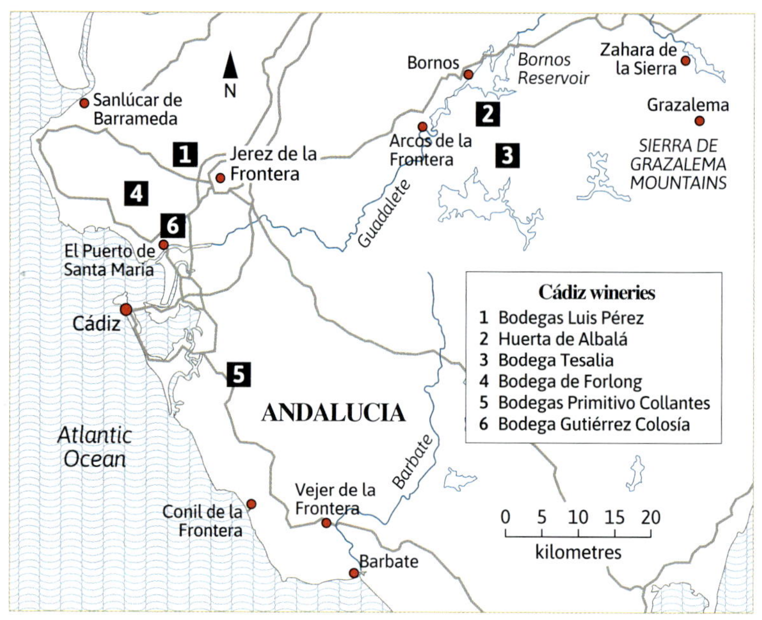

the vineyards and winery, and a wine tasting in the elegant main room of the farmhouse.

Set in some 90ha of rolling hills overlooking the Bornos reservoir (an area also known as 'the Tuscany of Spain') east of Arcos de la Frontera, the vineyards and winery of Huerta de Albalá (*huertadealbala.com*) are the life's work of the late Vicente Taberner Carsi, the result of his dream of creating a great Andalucían red wine. Visits (by appointment) are available Monday to Saturday and include tours of the vineyards, winery and cellar, and a sampling of the bodega's Barbazul and Taberner brand wines.

GETTING THERE

There are frequent flights from various destinations in Europe to Seville and Jerez de la Frontera with easy train connections to Cádiz.

Bodega Tesalia (*bodegatesalia.com*) was bought in 2007 by Richard and Francesca Golding when they came to Arcos from London and fell in love with the countryside on the edge of the Grazalema range. Their aim became to create high-quality red wines, now marketed under the names Tesalia, Arx and Iceni. As well as the vineyards and winery, the estate is used for breeding horses and general agriculture. Visits are run by daughter Natalia and are available by appointment only.

In 2007, Alejandro Narváez and Rocío Aspera took over the derelict estate formerly known as El Olivar de Forlón, just outside El Puerto de Santa María, and began the work of restoring the old vineyards, olive groves and orchards. Thus Bodega de Forlong (*bodegadeforlong.com*) was born. It is truly a labour of love – and especially a love and respect of terroir, in this case the local chalky albariza soil, with the wines expressing both the distinct and subtle differences between plots. The winery specialises in a delightful range of artisanal and ecological wines and visits are available upon request.

Casa de las Cuatro Torres

More than 150 years ago, brothers Primitivo and Tomás Collantes arrived in Chiclana de la Frontera from their northern home near Santander and set up their first winery, making wines from the local Palomino grapes. In 2014, with grandson Primitivo at the helm, Bodegas Primitivo Collantes (*bodegaprimitivocollantes.es*) expanded on its full range of of DO Jerez Sherry wines to include dry and semi-sweet still wines, and more recently has also recovered the indigenous Uva Rey variety. Due to its seaside location, towards Cádiz city itself, Chiclana benefits from the sea breezes to maintain the high quality of its biologically aged wines (aged under a layer of flor yeasts that protect it from the effects of oxygen). Visits to the bodega are offered Mondays to Fridays.

YOUR CÁDIZ ADDRESS BOOK

ACCOMMODATION

Casa de las Cuatro Torres
Traditional luxury in an 18th-century neoclassical-style merchant palace on the edge of Cádiz old town (pictured left), with splendid rooftop views of the port and across the city. **casadelascuatrotorres.com**

Hotel Argantonio
A perfect blend of classical Andaluz and Moorish architecture combined with modern comfort in a beautifully restored 18th-century house in the old city centre.
hotelargantonio.com

Parador Atlántico
A thoroughly modern version of Spain's famous Parador hotels, the Atlántico has all the luxuries you'd expect including a gym, spa and pool, in a perfect seafront setting with stunning views. **parador.es**

Mercado Central

RESTAURANTS

Casa Manteca
If you haven't been to Casa Manteca you haven't been to Cádiz. They've been serving traditional charcuterie and Sherry in their tiny bullfight memorabilia-festooned bar since 1953 and it's a natural starting point for pre-lunch or -dinner tapas. Their new freiduría (bar selling fried fish) across the street is also worth a visit. **facebook.com/tabernamanteca**

Ciclo
Chef Luis Callealta developed professionally alongside such great names as Angel León of Aponiente in El Puerto and Martín Berasategui near San Sebastián on Spain's north coast – both three-star Michelin. This new venture, near the cathedral, combines classic culinary excellence and professional service based on locally sourced quality products, with an excellent wine list to match.
ciclorestaurante.com

El Faro de Cádiz
Three generations of the Córdoba family and almost 60 years in the business have made El Faro de Cádiz one of the best known spots in the city for local fish and seafood, with tapas bar or restaurant options and an impressive wine list. **elfarodecadiz.com**

SHOPS & MARKETS

Alándalus Club
Just outside the old city, this is the place to go for locally sourced gourmet quality produce (from wines and conserves to olive oil and cheeses), cosmetics, and everything in between. You can order online, too.
alandalusclub.com

Mercado Central
Founded in 1837 with major renovations completed in 2009, the Central Market is the hub of commercial activity in the city. It has the full usual range of fresh produce, but the star of the show is the fabulous fish market, with plenty of small bars and shops lining the outside walls to refresh yourself and enjoy a snack. **mercadocentralcadiz.com**

Sherry Wines Shop
A great place to stock up on Sherry and vermouth from Bodegas Lustau in this charming stall located on the outer parade of the Central Market (stall No6). And of course you can try before you buy.
See Facebook

Hotel Meliá Madrid Princesa
METROPOLIS

MADRID

From traditional tabernas to trendy tapas bars and bustling markets, Madrid's food and drink culture is as expansive as the city itself. Get to know its varied neighbourhoods one by one.

Madrid can be an overwhelming experience, but it turns out that the trick to savouring Spain's capital is to take it in bite-sized pieces. It took me several visits over the past 30 years to finally warm to what I at first felt was an impersonal, sprawling metropolis.

Then I learned from friends who live here to take it barrio by barrio, and get to know the very different personalities of each neighbourhood. From trendy Malasaña to upscale Salamanca; from multicultural Lavapiés and hip Barrio de las Letras to traditional La Latina; each barrio has something unique to offer.

Happily, there has also been a surge of great new wine bars with a focus on small producers, both Spanish and international, plus chefs returning to the fundamentals of Spanish cooking. You can taste the past and present of Madrid in dishes such as the classic callos a la madrileña (tripe stew) at El Fogón de Trifón (see Facebook), or try innovative adaptations of traditional todo la vida (lifelong) favourites at Media Ración (*mediaracion.com*).

Local wines are also back in vogue. Vinos de Madrid acquired its DO status in 1990, and since then has been gaining a reputation for higher-quality, small-production wines. The DO is divided into three demarcated sub-regions: Arganda, Navalcarnero and San Martín, each of which makes its own distinct styles of wine.

Many wineries such as Las Moradas and Saavedra in San Martín have embraced wine tourism, while specialist single-parcel producers such as Comando G and Marañones in the mountainous Sierra de Gredos area are committed to reviving native Garnacha and Albillo vines.

Back in Madrid you'll find no shortage of food markets, but if you want to try a more castizo (rootsy or authentic) experience, then try the places where locals still go to shop and snack. Mercado Vallehermoso in the district of Chamberí combines a small-producers market of 22 stalls with a variety of wine and tapas bars. Meanwhile, in boho Barrio de las Letras, bordering on Lavapiés, Mercado de Antón Martín is a hip version of an unpretentious local market, with bars serving tapas, craft beer and wine dotted throughout.

The Matadero in Arganzuela, a former slaughterhouse that has been transformed into an international living arts centre, is a constantly changing creative space combining art, cinema, design and culture, with an artisan local food market on the last weekend of each month.

While San Isidro is probably the most traditional of Madrid's annual festivals, the fiestas of San Lorenzo in Lavapiés and La Paloma in La Latina are the lively ones, with a buzzy street-party vibe and parades. The latter features competitions of chotis, a style of traditional Madrileño music and dance with Bohemian roots.

Madrid is not only a moveable feast, it's a fast-moving one. Even long-time residents can find it difficult to keep up. It's impossible to see or taste it all at once, but with each visit you'll find new reasons to return.

OUR TOP 10

1. Angelita

Sommelier David Villalón and his brother Mario run this exceptional wine bar and bistro. More than 100 wines are available by the glass or half glass, perfect for pairing. Their menu is short and ever-changing, with organic veg from their mother Angelita's own garden outside the city. **madrid-angelita.es**

1. Angelita

2. Amano

With chef Javier Goya and somm-maître extraordinaire Fran Ramírez at the helm of this venture in the heart of Barrio de las Letras, great food, outstanding wines and knowledgeable service are guaranteed. Pop in for a mano tapas at the bar – to be eaten with your hands – or book a table in the restaurant. **amanomadrid.com**

3. Taberna Palo Cortado

Paqui Espinosa pays tribute to Sherry with a comprehensive selection of all styles, along with non-Sherry options, available by the glass and pairing perfectly with the seasonal Andalusian dishes on the menu. **tabernapalocortado.es**

4. Taberna de Pedro

Pedro García de la Navarra's charming tavern, a stone's throw from El Prado museum, is all about providing excellent products at reasonable prices. Traditional dishes are complemented by a splendid cheese board and a fantastic wine list created by García de la Navarra's brother Luis. **latabernadepedro.com**

5. Laskasa

Dream team César Martín (chef) and Marina Launay (front-of-house) have mastered that special blend of impeccable service and an intelligent seasonal menu. Almost everything is available in half-portions, including the wines by the glass. The all-day kitchen in this fun, trendy venue makes it a great spot for late lunches or early dinners. **lakasa.es**

6. Casa Gerardo

Cheese, please. This 80-year-old bodega boasts an impressive cheese list with an equally impressive selection of wines, including vermouths and Sherries. Great old-school atmosphere with friendly service. **See Facebook**

7. Bodegas Ricla

Founded in 1867, this tiny mother-and-son operation, just steps from the Plaza Mayor, is a classic stop for vermouth. It's a basic, no-nonsense tapas bar filled with memorabilia of a bygone age. Be sure to try the callos (tripe stew) or the famous meatballs. **See Facebook**

8. La Fisna Vinos

Working simultaneously as importers, distributors and retailers, Delia Baeza and Iñaki Gómez Legorburu have converted a rustic tavern in Lavapiés into a cosy wine bar and shop. Choose from a selection of unique French and Spanish wines from small producers, most sold exclusively here, at very affordable prices, along with delicious tapas. **See Facebook**

4. Taberna de Pedro

9. Taberna Verdejo

This small but perfectly formed tavern run by Marian Reguera and team has a menu full of tasty traditional cooking, a terrific wine list, a blackboard overflowing with Sherries by the glass and, best of all, the kind of service that makes you feel as welcome as you would be in your own home.

tabernaverdejo.com

10. Zalamero

Partners Ana Losada and David Moreno are the heart and soul of this lovely tavern. A cosy bar greets you, leading to a tasteful, minimalist dining area beyond. With 40-50 wines that change weekly and a market-based menu, it is simply spectacular.

zalamerotaberna.com

6. Casa Gerardo

RIOJA

The medieval hilltop town of Laguardia offers a stunning panorama of the region

This historic region is renowned for its vineyards, but the landscape is enhanced by ancient caves, monasteries and award-winning architecture. Follow Decanter's local guide to explore the area's bodegas, wines and other delights.

An old marketing slogan defined Rioja as 'The Land of a Thousand Wines'. Such a claim may sound exaggerated, especially if you think of the big-brand, cheap and cheerful crianzas lined up on supermarket shelves, but it feels less ridiculous once you discover the region's diversity, which goes well beyond its wines. Spanning 150km west to east along the Ebro river, Rioja is best visited by car. Driving offers you the chance to properly explore the vineyard-lined roads that traverse the region's seven river valleys, meander through hilltop towns set against two mountain ranges and admire the award-winning architecture and ancient monasteries straddling the Camino de Santiago pilgrims' trail.

If you'd prefer to relax and let others do all the work, local guides such as Riojatrek (*riojatrek.com*) or Amelí (*ameliriojatours.com*) can do the organising for you. Direct your efforts, instead, towards indulging in the rich food scene, whether that be dining at a top-class restaurant or a lower-key experience sampling tapas and a glass of local vino in Logroño's famous Calle Laurel or the busy Tastavin wine bar nearby on Calle San Juan.

Whether you choose the buzz of the city or the peaceful atmosphere of pretty villages such as Samaniego or Briñas for your stay, spring and early summer are probably the best seasons for travelling to the region. Alternatively, plan your trip to coincide with the grape harvest in September and October.

TOP Sajazarra is one of Spain's prettiest villages.
BELOW Bodegas Ysios sits against the backdrop of the Sierra Cantabria.

'Most of Rioja's historic caves are now used as private leisure spaces, but a few still operate as fully fledged wineries'

As there are no great distances involved, it is easy to get to charming villages off the beaten track such as Labraza, in the far east of Rioja Alavesa. Heading west, Sajazarra has a beautifully preserved castle and is among the prettiest villages in Spain. In 1899, the remains of its medieval walls were witness to the first outbreak of phylloxera in Rioja, in a local vineyard.

A full agenda

To enjoy a bird's-eye view of the region, there's nothing like driving up the A-2124 on a clear day, towards the summit of the Sierra de Cantabria mountains. Stop at the Balcón de la Rioja lookout for panoramic views of the vineyards and villages on both sides of the river.

If this view is key to understanding the geography of Rioja, then the medieval hilltop town of Laguardia offers a different outlook – a stunning panorama of vines and wineries set against the backdrop of the mountains. Bodegas Ysios (*bodegasysios.com*) and its wavy metal roof provide the perfect Instagram-ready shot, especially if you manage to catch the clouds sliding over the crest of the Sierra Cantabria down the slope for that Foehn effect. Daily 'winecar' tours, vineyard walks, and tutored tastings can be booked online.

If you'd prefer to enjoy this view from a chaise longue, glass of wine in hand, the place to go is the stylish new wine bar outside the gates of Bodegas Javier San Pedro Ortega (*bodegasjavier sanpedro.com*). Javier, a fifth-generation

MY PERFECT DAY IN RIOJA

MORNING

After an invigorating sleep and breakfast at the tranquil **Palacio Tondón** *(marriott.com)* in Briñas, complete with views of the Ebro river and the vines of Viña Tondonia, grab your walking shoes and hop in the car for a 10-minute drive to **Remelluri** *(remelluri.com)*. The estate is not open for tours, but visitors are welcome to enjoy a self-guided walk through its organic vineyards, taking a peek at the 10th-century necropolis and ancient stone lagar (wine press), carved from a huge granite boulder among the vines. A short drive southwest is Haro and its **Barrio de la Estación** (railway station district). You'll be spoilt for choice, but regardless of whether you go to **CVNE** *(cvne.com)*, with its ageing cellar designed by architect Gustave Eiffel, or to **Gómez Cruzado** *(gomezcruzado.com)*, the smallest of the seven wineries, fun is guaranteed.

Marqués de Riscal Hotel, designed by Frank Gehry

LUNCH & AFTERNOON

If you want to combine a winery tour with a tasting and a bite to eat, **Bodegas Roda** *(roda.es)* in Barrio de la Estación offers just that with its lovely balcony overlooking the river. Alternatively, you can drive for 30 minutes to Laguardia with a stopover at the San Vicente lookout in Elciego to take in the views of the **Marqués de Riscal Hotel** and its multicoloured roof standing in contrast to the surrounding landscape. Once in Laguardia, head to **Amelibia** *(restauranteamelibia.com)* outside the city

Views of the Ebro from Palacio Tondón

walls, for a delicious meal prepared with seasonal ingredients and a wine list featuring small Rioja producers.

Wandering through the streets of Laguardia is a joyful experience, particularly outside the peak holiday season. Another nearby hilltop town, San Vicente de la Sonsierra is also dotted with imposing buildings such as the **Bodega Teodoro Ruiz Monge** *(bodegateodororuizmonge.com)*, which is run by an artisanal grower who will happily show you around the family's centenarian cellars and vineyards.

Haro's Barrio de la Estación is home to seven big wineries

EVENING

For dinner, drive back to Haro, but this time head to the old town. In one of its many squares, you will find **Nublo** *(nublorestaurant.com)*, a restaurant set in a 16th-century palace where all the food is cooked over open flames.

Nublo restaurant

grower, makes a diverse range of wines, from aromatic whites to serious single-vineyard reds. Daily visits to the winery can be booked online, or enjoy a 'quick guided tasting' (with tapas) in the bar.

Architecture and wine are deeply entwined in Rioja, from iconic modern buildings such as Frank Gehry's Hotel Marqués de Riscal in Elciego *(marquesderiscal.com)* to traditional wineries including Conde de los Andes in Ollauri *(bodegasollauri.com)* with its intricate maze of underground cellars housing dozens of historical vintages.

Few people know more about this than architect-turned-winemaker Javier Arizcuren. As well as restoring Conde de los Andes and building modern wineries such as Finca de los Arandinos *(fincadelosarandinos.com)*, he makes a handful of quality wines from his family vineyards in Rioja Oriental. Arizcuren may one day restore his ancestors' cellar in Quel's 18th-century bodega district but, for now, he works in a garage winery – the only one in the centre of Logroño that is open to visitors. Check the website for details *(arizcurenvinos.com)*.

GETTING THERE

Logroño is a 90-minute drive from Bilbao airport and Madrid is less than four hours away. There are also direct links by train and bus.

Immerse yourself in Rioja

Like those in Quel, most of Rioja's historic caves are now used as private leisure spaces, but a few still operate as fully fledged wineries. A typical example is Bodegas Lecea *(bodegaslecea.com)*, one of 300 underground cellars created in the 16th century in San Asensio. The Lecea family makes genuinely traditional wine, including its carbonic maceration red Corazón de Lago. Anyone visiting at the end of the harvest is invited to join in the fun and tread the Tempranillo grapes with their feet in the old stone press. Otherwise, bookable tours are available: a standard 'English' option (Monday to

Samaniego church towers over the old village

Raucous partying at the Batalla del Vino

Friday only); a daily 'premium tour', which offers visitors the chance to help out in the vineyard (sustenance included); and a 'gastronomic visit' complete with a Rioja lunch.

For a touch of glamour, Haro's famous Barrio de la Estación (railway station district) hosts the biennial Haro Station Wine Experience. No visit to Rioja is complete without going to the Barrio, with its seven centuries-old wineries, but if you plan your trip for the summer, you will have the chance to visit six of the prestigious bodegas in the barrio – La Rioja Alta, Muga, CVNE, Bilbaínas, Roda and Gómez Cruzado – and taste their full range of wines paired with pintxos prepared by some of the region's famous chefs. Tickets are available to buy from the Haro Station website *(barrioestacion.com)*.

Note that López de Heredia Viña Tondonia *(lopezdeheredia.com)*, the Barrio's oldest neighbour, is no longer open to visitors, but you can still peek into its decanter-shaped store built by Zaha Hadid and buy its wines – apart from the cult releases.

Staged not far from Rioja's golden mile, the Batalla del Vino or 'Wine Battle' *(harowinefight.com)* is a raucous annual party which involves hundreds of people happily throwing wine at each other. It always takes place on Saint Peter's Day, which is on 29 June every year, with the combatants wearing traditional red and white outfits.

Old meets new at the Hospedería de los Parajes

YOUR RIOJA ADDRESS BOOK

ACCOMMODATION

Hospedería de los Parajes Located in Laguardia, this hotel has spacious rooms, a bar in a 15th-century cellar and two restaurants offering generous, traditional dishes. **hospederiadelosparajes.com**

Hotel Calle Mayor Logroño Set in the town's historic centre, this small hotel is handy if you want to enjoy the tapas and wine bars a few blocks away without having to drive. Helpful staff and excellent breakfast. **hotelcallemayor.com**

Palacio de Samaniego Luxury rooms in a 17th-century palace owned by the Rothschild family. Expect an outdoor lap pool, creative dishes in the restaurant and the chance to visit the Macán estate, co-owned by Vega Sicilia. **palaciodesamaniego.com**

Michelin-starred dining at Venta Moncalvillo

RESTAURANTS

Héctor Oribe Family-owned, this restaurant offers one of the best lunchtime menus in Rioja Alavesa, hence its popularity among wine producers. Oribe's traditional food is matched by a cellar containing more than 100 wines. **hectororibe.es**

Ikaro Run by a young Spanish-Ecuadorian couple, this 'fine dining' one-star Michelin, but moderately priced, restaurant offers a creative vision of local gastronomy combined with some fusion dishes. **restauranteikaro.com**

Venta Moncalvillo A one-star Michelin restaurant owned by brothers Ignacio and Carlos Echapresto. Many of the ingredients are sourced from their vegetable garden, visible from the dining room. The wine cellar boasts more than 1,300 cuvées, with a focus on Rioja. **ventamoncalvillo.com/web**

PLACES TO VISIT

Erroiz Few people are aware that Rioja also produces some excellent olive oil. At this mill in Lanciego, you can see old groves and sample extra-virgin oils. **erroiz.eus**

Vivanco Museum of Wine Culture Next to the Vivanco winery in Briones, this museum houses a grapevine garden with more than 220 varieties, a vast wine-themed art collection with works by the likes of Picasso and Warhol, and one of the world's finest collections of corkscrews. **vivancoculturadevino.es**

THE ALGARVE

Its year-round sun and dramatic coastline mean it's already a much-loved beach holiday destination. But this southern Portuguese hotspot is upping its game, providing plenty to tempt wine lovers inland for a taste of something different.

Spectacular limestone cliffs and Arcos Naturais double sea arch, as seen from the popular Praia da Marinha beach

When the Romans reached the southwestern point of the Algarve, they thought it was the end of the world, where the waters of the ocean boiled at sunset. Yet despite the impending sense of doom (or perhaps because of it), they planted vines in the region, finding the temperate climate and fertile terroir a nirvana for wine-growing.

Fast forward to present times and it isn't just the grapes that relish soaking up the rays, as beach lovers, walkers, cyclists, golfers and water sports enthusiasts all bask in their share of a reputed annual 300 days of sunshine. At first, the coastal vineyards lost out to the consequent package holiday boom of lucrative hotels and seaside resorts developed from the 1960s onwards. But in a land deeply rooted in wine-growing traditions, artisanal viticulture is re-emerging in a flourishing revival of indigenous red and white grape varieties, especially the revered Negra Mole (meaning 'black soft').

As enterprising estate owners become increasingly recognised for award-winning results, a new set of adventurous wine tourism thrill-seekers is fast being drawn to Portugal's south. This final frontier of western Europe still has so many grapes yet to be tasted and explored, along with the region's rich gastronomy, culture and dramatic landscapes.

Many vineyards are concentrated inland in what is known as the Barrocal, bookended between the Atlantic and the rugged uplands with soils of sand, limestone, clay, shale and alluvium. Producers are working hard to safeguard the distinctive properties of the Algarve's wines, particularly in the region's four DOCs (or DOPs, west to east): Lagos, Portimão, Lagoa and Tavira.

Making memories

Warm, velvety reds include Negra Mole, Castelão and Trincadeira, while top DO white wines such as Arinto, Malvasia Fina and Crato Branco (Síria) taste delicate and smooth. A

'This final frontier of western Europe still has so many grapes yet to be tasted and explored'

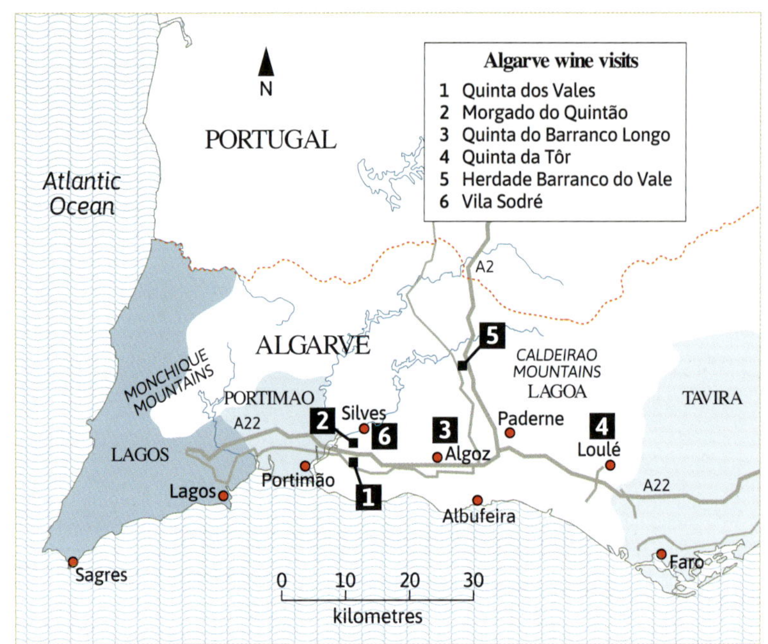

significant number of Algarve Vinho Regional wines also allow producers even more versatility, with sparkling and late-harvest wines recently released.

Although harvest can start as soon as late July, early September is a good time to catch it. At family-run Quinta dos Vales *(quintadosvales.eu)* in Lagoa, the enterprising Winemaker Experience package includes the opportunity to rent or even own a vineyard along with the facilities to make your own wine at this colourful, art-loving estate (pictured right) decorated with voluptuous sculptures amid rolling vines, stylish villas, tennis courts and pools (private and shared).

GETTING THERE

Low-cost airlines from regional UK airports fly to Faro's international airport. Rent a car on arrival for the approximate hour-long drive to the heart of the Algarve wine region. For more information go to visitalgarve.pt

For a slightly less ambitious start to a career in winemaking, the three-hour bottle blending workshop allows you to handpick grapes before sitting down with experts to glean tips on tasting and combine different proportions of native and international varieties. You can create your own blend (I chose 40% Touriga Nacional, 40% Cabernet Sauvignon and 20% Aragonês), before hand-corking and labelling bottles, mine since set aside for an impending special birthday...

Vila Vita Parc, Porches (see p98)

A private bash at Morgado do Quintão *(morgadodoquintao.pt)* would be idyllic, and tours, tastings, lunches, weddings and authentic farm stays in vineyard cottages are popular here. To spend time sitting in the shade of a 2,000-year-old olive tree overlooking Negra Mole, Crato Branco and Castelão vines at a farmer's table of Algarvian delicacies paired with excellent estate wines is truly memorable.

First founded in 1810 by the Count of Silves and located in a wine-rich region sandwiched between the Monchique mountains and Lagoa, this historic winery is still owned by the fourth generation of the original family. Their passion for preserving the inherited indigenous grapes is undiminished, championing centuries-old styles such as palhete and clarete – both light, pale-hued reds, the former made by mixing with up to 15% white grapes. Yet a modern vision to return to organic

Quinta dos Vales

MY PERFECT DAY IN THE ALGARVE

MORNING

An al fresco breakfast of a still-warm pastel de nata with a freshly squeezed orange juice and hot, strong coffee is, in my books, obligatory in the Algarve. The spectacular, golden cliff coastline is best appreciated from the sea, so set sail from Albufeira marina with **AlgarExperience** *(algarexperience.com)* on a thrilling two-hour speedboat tour, with fun commentary on the bays, caves, dolphins, lighthouses and celebrity homes.

AlgarExperience

LUNCH & AFTERNOON

Keep the adrenaline going with an **Extremo Ambiente** *(extremoambiente.pt)* Jeep Safari to the breathtaking Monchique mountains – you can be picked up from Albufeira, and opt to stop off for an atmospheric lunch and wine tasting near historic Silves, at fascinating family cellar **Vila Sodré** *(residencialvilasodre.pt)*.

It's a treasure trove of fortified and still wines dating back as early as 1890. Knowledgeable and enthusiastic sommelier Andre will pour glass after glass of Portuguese wines, from a local Al-Mudd Negra Mole Rosé 2020 to a Messias 1966 vintage Port from the Douro, at tables laden with Algarvian cheeses, homemade bread, sausages, olives, sardine pâté, carob cake and almond tart. On the steep, winding uphill climb, fortify yourself with the local brandy Medronho and Melosa liqueur, made from honey and wine, at a mountain village cave before enjoying one of the most stunning panoramic views at sunset.

Restaurante Veneza

EVENING

For a memorable menu paired with the best vintages, **Restaurante Veneza** *(restauranteveneza.com)* is in Paderne, an 11km drive inland from Albufeira. You'll be surrounded by 2,000 different fine wines, on sale by the glass or to take away by the bottle. Each delicious wine and dish is described and served by knowledgeable staff in this friendly, family-run establishment which started out as grandfather Manel's grocery store. If you don't want to drive, dinner here can also be added to the action and wine-packed Extremo Ambiente Jeep Safari.

production has resulted in prize-winning wines which, since first being produced in 2016, have already won annual awards from Revista de Vinhos magazine for 'Best Wine in Portugal: Algarve' in 2019 and 2020.

Perhaps it's no surprise that the wines at Quinta do Barranco Longo *(quintadobarrancolongo.com)* have reached the top division, as owner Rui Virgínia's two footballing sons have both had stints in the English Premier League! Located in rural Algoz, the winery's use of modern technologies combined with innovative winemaking methods have resulted in fine wines from blends of native and international grape varieties, such as the vibrant Grande Escolha (Arinto, Chardonnay, Encruzado) or refreshing Aragonês-Touriga Nacional Rosé. This year, a new tasting room will open with a rooftop restaurant, which promises to be an incredible place to drink in both spectacular wines and scenery.

Ancient & Modern

You can lap up grape views while taking a dip in the infinity pool during a tasting tour at Quinta da Tôr *(quintadator.com)*, now owned by Mário Santos, local-lad-made-winemaker, who

YOUR ALGARVE ADDRESS BOOK

ACCOMMODATION

3HB FARO

The first five-star boutique property in the Algarve capital with a rooftop infinity pool, cocktail bar and gourmet restaurant with panoramic views. Only 15 minutes by taxi from Faro airport, there are stylish rooms and a modern spa. **3hb.com/home/3hb-faro**

São Rafael Atlântico

A contemporary five-star hotel in Albufeira with spacious rooms and a spa as well as direct beach access and three outdoor pools in its tropical gardens. Breakfast can be enjoyed on the terrace, with pancakes and other treats made freshly for you. **saorafaelatlantico.com**

O Pátio

Vila Vita Parc

This five-star resort on the coast in Porches offers private tastings in the exclusive wine cellar or dinner in 10 different restaurants, including Michelin two-star Ocean. Many of the rooms, apartments and villas have sea views; others look out over the lush gardens with fountains, pools and resident swans. **vilavitaparc.com**

RESTAURANTS

O Pátio

Head to coastal Carvoeiro for delicious seafood. This is one of the oldest and best restaurants in the area. The original water well has been converted into a wine cellar. **opatiocarvoeiro.com**

Restaurante Ramires

Don't miss another signature Algarve dish, piri-piri chicken. This friendly, family-owned restaurant in Guia, Albufeira, has been serving it since 1964. **restauranteramires.com**

Tertúlia Algarvia

Doubles as a restaurant and cooking school in Faro. Learn to cook a traditional seafood stew in the iconic, clam-shaped cataplana, enjoying the fruits of your labour afterwards on the sunny terrace. **tertulia-algarvia.pt**

SHOPS & LEISURE

Adega-Museu de Odeceixe

The winery museum near Aljezur is housed in an early 20th-century winery. Visitors can discover the history of wine production through antique tools of the trade. **visitalgarve.pt/equipamento/5497/adega-museu-de-odeceixe**

Mercados de Olhão

Head to this vibrant market in Olhão with your friendly Portugal4U guide before catching a water taxi to Culatra Island in the Ria Formosa lagoon, where you learn about oyster harvesting. Wine tours are also available. **pt4u.pt/tours**

Seven Hanging Valleys Trail

Hike the seven-mile track from Praia da Marinha to Praia do Vale de Centeanes in Carvoeiro above some of the world's best beaches with cliff-edge viewpoints. For longer routes, walkers and cyclists should follow the Coastal Ecovia (along the whole southern Algarve coast) or the inland Via Algarviana, as many lead to wineries. **walkalgarve.com**

'In a land deeply rooted in wine-growing traditions, artisanal viticulture is re-emerging'

as a boy used to go swimming in the nearby river. The historic Roman bridge that leads to Tôr village, just north of Loulé, adorns the labels of the estate wines, which are known for their unusually high alcohol content, such as the robust 17% Syrah.

Surrounded by cork, carob, olive and pine trees, Herdade Barranco do Vale *(hbv.pt)* offers stunning panoramas of the Caldeirão mountains when you take a tour of the estate on foot or by tractor. Owner Ana Chaves still tends the Negra Mole vines that were first planted here by her grandfather in the 1960s, as well as 20-year-old Aragonês and Castelão vines, but the latest family generation has now also introduced first-class white grape varieties such as Antão Vaz, Arinto and Alvarinho.

While the fine wines in northern Portugal have long been internationally revered, the Algarve wine-producing region is only now coming of age. It's an exciting time for discerning wine lovers who will undoubtedly relish heading to the 'ends of the earth' for the dawning of a new wine era – and perhaps stock up on some bargain bottles in the process.

RIGHT The stunning sea cave at Benagil.

UNITED STATES & CANADA

There's more to North American fine wine than Napa. Wend your way at a leisurely pace along Mendocino's sleepy coastline or, beyond California, explore Long Island, Walla Walla or Canada's Vancouver Island. And for the really adventurous, Idaho, Colorado, Texas, Michigan and North Carolina are all well off the beaten wine tourist track, with unfamiliar wines just waiting to be discovered …

CHARLESTON

Putting together a hitlist of US wine destinations? This South Carolina port city probably wouldn't spring to mind – but it should. Charleston has a long association with the wine trade, and its bustling and innovative scene draws professionals and tourists alike.

Traditional houses alongside Colonial Lake in historic Charleston

Perhaps surprisingly for a city in the American South, where Vitis vinifera suffers in the summer heat and humidity, Charleston has cultivated a long relationship with wine. Founded by English colonists in 1670, Charleston ('Charles Town') evolved into a key East Coast port through which volumes of Madeira flowed.

In the early 1800s, social clubs dedicated to fortified wine popped up. During the American Civil War, the South Carolina Jockey Club, a racing organisation, hid its valuable Madeira stash in the basement of a Columbia asylum on news of an advancing Union Army.

Traces of this vinous legacy remain as Charleston balances the preservation of its heritage against reinvention for a modern world. Today, the 'Holy City' is enjoying a new wine renaissance, with numerous wine-centric restaurants, wine shops and bars opening regularly over the last several years.

During the pandemic, I relocated from New York to this small coastal city, seeking a better quality of life. Other wine and restaurant professionals had the same idea. Many moved temporarily, fell in love, and then stayed, contributing new dimensions of expertise to a region famous for its Lowcountry cuisine and Southern hospitality. Charleston's interest in wine had already been growing, however, says Kellie Holmes, a restaurant wine consultant who relocated from San Francisco in 2011. Over the past decade, she's witnessed countless changes.

'I've seen smaller distributors enter the market, smaller wine shops and bars open, more natural wine and emphasis on sustainability, plus more wine lovers interested in categories like pét-nat and grower Champagne,' she says.

The Charleston Wine & Food Festival (charlestonwineandfood.com), founded in 2005 by community leaders, helped catapult the historic city into the spotlight. The five-day event takes place every March, with the 2024 event scheduled for 6-10 March. Big-name chefs, sommeliers, bartenders and even wine writers started attending, generating a buzz that fuelled itself and contributed directly to the growth of the hospitality sector. 'It's an unexpected but pleasant surprise for travellers to learn that Charleston has such a vibrant wine scene,' says Holmes, 'thanks to the young professionals who have been working here, staying put, and turning into business owners who are passionate about wine,' she says.

YOUR CHARLESTON ADDRESS BOOK

Bar Rollins

194 Jackson Street, Charleston SC 29403

What started as a pop-up wine bar now has a permanent home. Bar Rollins, a natural-wine spot on Charleston's East Side, opened in June 2022 with a selection of whites, rosés and reds by the glass, including Chenin Blanc, Fer Servadou and Grolleau. Tapping into the zero-alcohol trend, it also offers several non-alcoholic options.

Before opening, owners Chris Rollins and Jess Vande Werken spent a few years working hard on social media to announce pop-up locations and collaborations with local chefs, including FIG's Jason Stanhope, to generate buzz for their concept. Their 'wine dive bar' occupies a classic Charleston single house (long and narrow, just one room in width), featuring bar and lounge seating inside and a patio and porch at the back. Guests get a verbal hug before entering from the 'Bar Rollins loves you' inscription on the sidewalk outside. Mismatched bistro furniture, a communal wooden bench, exposed brickwork and original fireplace details give the space a schoolhouse vibe. Open Wednesday to Monday, 4pm-10pm. No reservation required. **barrollins.com**

Chris Rollins (right), with his Bar Rollins team

Miles White (top) and Femi Oyediran, Graft Wine Shop

Graft Wine Shop

700B King St, Suite B, Charleston, SC 29403

Opened in 2018 by wine and music lovers Femi Oyediran and Miles White, Graft has become synonymous with eclectic, minimalist wines, served and sold in a contemporary space. For wine drinkers visiting Charleston, Graft usually lands at the top of the must-visit list. Success, however, was a decade in the making. The pair met 10 years prior while working at local fine-dining institution Charleston Grill under the tutelage of the city's wine maestro and connector Rick Rubel.

Both Oyediran and White have gone on to win accolades for their talents but have never strayed from the goal of making wine accessible and fun. They've welcomed guest chefs for hip-hop events, including at Graft's yearly contribution to the Charleston Wine & Food Festival. As music junkies, White and Oyediran release a monthly playlist to match the seasonal wine mood. The bar menu covers categories such as 'White Wines Are Cool' and '#Reds4Dayz'. Cured meats, conservas (speciality tinned fish and seafood) and cheeses provide snacks. Wine shop open daily 12pm-10pm; bar open Monday to Friday 4.30pm-10pm, Saturday and Sunday 12pm-10pm. **graftchs.com**

Herd Provisions

106 Grove Street, Charleston, SC 29403

The relaxed energy of this two-storey modern farmhouse restaurant belies the level of detail paid to the wine list. At first glance, burgers, poutine, pumpkin ravioli and fancy cheese toasties sound like comfort food in need of comfort wine. Read a little deeper into the menu, however, and a theme emerges: globally inspired dishes using sustainably and locally sourced ingredients, which form the foundation of the kitchen. The meat hails from Leaping Waters Farm, a 240ha pasture in Shawsville, southwest Virginia (which is also owned by the restaurant's owner) and is butchered in-house at Herd. As such, the menu requires a broad, characterful wine list that matches the restaurant's 'farm-to-table' ethos.

Curated by restaurant wine consultant Kellie Holmes, the wine list spans regions from California to Chablis, with particular attention paid to cooler-climate areas such as Alto Adige and Austria. Where she can, Holmes looks for sustainable, organic and biodynamic producers that also match her philosophy (she is a board member of Slow Food Charleston's 'Slow Wine' arm). Open Monday to Thursday 3pm-10pm, Friday and Saturday 11am-10pm, with happy hour 3pm-5pm daily except Sunday (when closed). Reservations are not required but are appreciated at dinner. **herdprovisions.com**

ABOVE RIGHT Brown butter scallops, asparagus coulis, pine nuts and oven-dried tomato, SAVI (pictured right)

SAVI Cucina + Wine Bar

1324 Theater Drive, Mount Pleasant, SC 29464

Located in Mount Pleasant, SAVI Cucina takes the wine experience seriously, having assembled a team with more wine credibility than entire American towns. Co-owner Ty Raju and SAVI Society Wine Club director Makayla Woodford have recently added beverage director Emerson Ewald to the team. Ewald spent six years as a certified sommelier at The Modern (at New York's MoMa) and as lead sommelier at Verōnika in New York City.

Together they have created an Italian list to cater for neophytes and experts alike, with bottles to match the house-made pasta, cheeses, sauces and dry-aged steaks. Guests can have a drink at the bar, enjoy dinner inside amid Italian coastal decor, or relax outside with a flight of Italian wines or a weighty, collector-worthy bottle. The wine club keeps locals engaged with special group tastings and private wine dinners. Open Monday to Saturday, with an aperitivo hour from 4pm-6pm; dinner service begins at 5pm. Reservations are preferred but not required. **savicucina.com**

Stems & Skins

Stems & Skins

1070 East Montague Ave #B, North Charleston, SC 29405

Always in search of New York neighbourhood parallels, realtors have called North Charleston's Park Circle the next Williamsburg. While this is a stretch of the imagination, the area is cheaper than downtown Charleston, thus attracting a younger clientele. Stems & Skins, with its tagline 'Fresh and Freaky ferments', stepped in to serve that younger audience with a neighbourhood natural-wine bar.

Owners Justin Croxall and Matt Tunstall (formerly of the renowned Husk restaurant in central Charleston) hold weeknight specials such as Aperitivo Hour, Vinyl Night and Meatball Night to cultivate a local vibe. A roster of minimalist wines, classic cocktails and craft beers paired with Mediterranean-style plates including charcuterie and conservas earned Stems & Skins a James Beard semifinalist nod for Outstanding Wine Program. In December 2021, the pair debuted a seafood-based eatery called Three Sirens on the same block with wine selections of a similar ethos. Open Monday-Friday 4pm-12am, Saturday and Sunday 2pm-12am. No reservation required. **stemsandskins.com**

The Tippling House

Charleston bars & restaurants

1. Bar Rollins
2. Graft Wine Shop
3. Herd Provisions
4. SAVI Cucina & Wine Bar
5. Stems & Skins
6. The Obstinate Daughter
7. The Restaurant & Bar at Zero George
8. The Tippling House
9. Vern's

NORTH CHARLESTON
DANIEL ISLAND
Wando River
CHARLESTON
DRUM ISLAND
Wappoo Creek
JAMES ISLAND
Charleston Harbour
SULLIVANS ISLAND
N
0 1 2 kilometres

The Tippling House

221 Coming Street, Charleston SC 29403

Sommelier Matthew Conway, formerly of Marc Forgione in New York, and his now-wife Carissa Hernandez, fled south when Covid shuttered the city. A stint with family in Charleston turned into a permanent relocation with the October 2021 opening of Conway's wine bar, The Tippling House. Conway spent much of the pandemic renovating the ground floor of a Charleston single house to prepare the intimate space.

The wine list, specials and menu have grown over the past year. For the recently debuted 'dinner parties' hosted on Wednesdays and Thursdays, Conway has partnered with chef Sean Clinton to serve a prix-fixe menu with one seating at 7pm. Other offers include a new by-the-glass pour on Fridays and winemaker tastings once a month. Everything on the daily wine list is available by the half-bottle, and Conway has assembled one of the best Syrah and 'reserve' lists in town. Open Tuesday to Saturday, 4pm-10pm. Other than for dinner parties, The Tippling House doesn't take reservations, so arrive early, especially for happy hour. **thetipplinghousechs.com**

Chef Vinson Petrillo leads a masterclass at The Restaurant & Bar at Zero George

Jacques Larson, owner-chef, The Obstinate Daughter

The Restaurant & Bar at Zero George

0 George Street, Charleston, SC 29401

Tucked away in Charleston's historic Ansonborough neighbourhood, a visit to The Restaurant & Bar at the Zero George Hotel evokes dinner at a friend's meticulously renovated 19th-century home. Gas lanterns flank the entrance off the leafy street, inviting diners into a lush setting of ferns and palms fit for the fine, inventive food of chef Vinson Petrillo. Beverage director Megan Mina curates the wine list with as much care given to her $50 selections as her older vintages from Burgundy and Bordeaux, some commanding $2,500.

In addition to the dinner tasting menu, guests can now book a caviar tasting at The Caviar Bar. While seated on the terrace, they can also access Mina's wine list, along with selected Champagnes by the glass and bottle, to pair with exquisite briny bites. Petrillo's previous experience as chef de cuisine at Caviar Russe in New York City shows in his refined dishes integrated with Regalis and Regiis Ova caviars.

Anyone who wants to delve deeper into the wine and food program can book an afternoon at the restaurant's cookery school. Petrillo and sous chef Chavis lead students through a multi-course meal preparation with wine pairings. The restaurant is open Tuesday to Sunday 6pm-10pm; bar open daily 5pm-11pm. Cooking classes Saturday and Sunday, 11am-1pm. Reservations essential. **zerogeorge.com**

The Obstinate Daughter

2063 Middle St, Sullivan's Island, SC 29482

Though open on Sullivan's Island since 2014, this eclectic Southern restaurant with a Mediterranean twist grows more popular each year. The name references the island's 'obstinance' during the Revolutionary War when American Patriots defending Fort Sullivan saved Charleston from British capture in 1776. The menu, devised by owner-chef Jacques Larson, changes seasonally with the ingredients sourced from Lowcountry farmers and fishermen.

The wine list looks to Italy, especially Italian whites, given the breadth of fresh seafood featured on the menu. There's a good range of crisp, savoury and mineral-driven expressions from the north to explore, whether Piedmont or Alto Adige. It's no coincidence that wine distributors like to have meetings here – languorous lunches over a platter of glittering oysters and a bottle of Gavi feel like a holiday even when talking business.

Larson takes pride in sustainability, having earned the two-storey spot Green Restaurant certification, and he extends that philosophy to wine selections where possible. Open daily 11am-3pm and 4pm-10pm (Friday and Saturday until 11pm). Dinner reservations are essential; lunch gets busy, but walk-ins are accepted. **theobstinatedaughter.com**

Vern's

41 Bogard St A, Charleston, SC, 29403

Newcomer Vern's, opened in July 2022 by former staff of the city's shuttered institution McCrady's, boasts the most exciting wine list to debut in Charleston last year. This modern American bistro, occupying a windowed corner spot downtown, radiates a convivial, intimate ambience and is ripe for becoming a neighbourhood joint with a twist – an ambitious, fun but philosophical wine list. Bethany Heinze, co-owner and operations/beverage director, together with her husband and executive chef Daniel 'Dano' Heinze, designed the food and wine selections around the local, seasonal and sustainable guiding principles. Though you may have heard that line before, you're unlikely to have seen a wine list as rich and quirky with the 'pioneers, renegades and old-school producers' of low-intervention farming and winemaking.

For a program of almost 100 selections, Bethany uses at least eight distributors to capture the breadth and singularity of labels she hopes will keep guests curious and coming back, whether for Aligoté from Sylvain Pataille or one of four skin-contact wines she rotates. 'I admit I was surprised with the level of demand for orange wine given the perception of Charleston being more traditional in taste,' she says. Clearly, change is afoot in this southern coastal city. Open Thursday to Monday, 5pm-10pm (closing 8pm on Sunday), brunch on Saturday and Sunday, 12pm-2pm. Reservations essential. **vernschs.com**

LONG ISLAND, NEW YORK

To the east of New York City, the wine producers of Suffolk County are helping to reshape the island's long history, with the fresh sea air, boutique lodgings and hyper-local cuisine of this much-loved coastal stretch now proving a big draw for the serious wine traveller.

The Shelter Island Heights community and yacht club with a view across the bay towards Greenport

'As Swiss wine stays in Switzerland, so Long Island's wines stay put, too'

Battling traffic on the Long Island Expressway (LIE) towards wine country, about 120km east of Manhattan, you'll find references to the layered history and various peoples who have inhabited the land in passing village and landmark names – Native Americans, Dutch and English settlers all left their imprint on New York's Long Island. Visual cues also give away proximity to wine country. You'll know you're close to the promised land when the multi-lane highway falls away, suburbia thins out into flat, open farmland, and green signs marked with arrows and a cluster of white grapes pop up pointing 'that-a-way'.

Wine production begins where the island splits, or forks into two peninsulas, at the town of Riverhead on Peconic Bay. The warmer, temperate climate (in comparison with New York state's other wine regions) creates richer, more full-bodied reds from mostly Bordeaux varieties. The vibe differs, too. Here, weathered cedar-shingled homes frame moody grey water views, while clam shacks and ephemeral summer farm stands, bountiful with stone fruits and berry pies, dot country lanes.

Long Island has three American Viticultural Areas (AVAs): North Fork of Long Island, The Hamptons, and a broader Long Island AVA, with 57 wine producers across all three. The North Fork AVA, established in 1986, remains the most relevant for wine tourists. It spans approximately 400km^2 across the northern peninsula and includes the Robins and Shelter Islands.

The South Fork, defined by its enviable trim of sandy beaches along the Atlantic ocean, developed into a glitzy real estate enclave called The Hamptons. Due to land prices and labour costs, only a few wineries operate there. The North Fork, hemmed in by the Long Island Sound to the north and Peconic Bay to the south, evolved into a sleepier agricultural hub.

Wine tourism has always factored into the region's development. Most wineries, being small or medium-sized, sell locally or regionally. Many move special cuvées, experimental

'Seated tastings in private tents, charming patios and flower-filled gardens have flourished'

Paumanok estate and buildings

projects and small-production varietal bottlings through direct-to-consumer programmes, from walk-ins to wine clubs. As Swiss wine stays in Switzerland, so Long Island's wines tend to stay put, too.

Maritime profile

In 1973, Louisa and Alex Hargrave planted Long Island's first vines. Vineyards started sprouting in earnest in the 1980s and '90s, replacing farms with rows of French grapevines. Well-drained sandy loam soils suit the flat terrain. The moderate climate invites a range of varieties. As a New World region, flexibility is the rule rather than the exception. However, producers must adhere to one regulation: 85% of the fruit bottled and labelled with an AVA must be grown within the borders of said AVA.

Surrounded by water, the vineyards here are both blessed and vexed by a mercurial maritime climate. A fickle spring, followed by spikes and drops in the summer heat, heavy rainfall and high humidity, combined with a long growing season, bring mould, mildew and pests. These challenges have become more acute with climate change. Long Island offers an honest accounting of the vintage in every bottle.

On the wine route

Despite the proximity to New York City and the uber-wealthy second homeowners of The Hamptons, Long Island's wine tourism industry once chased the low-hanging fruit of party buses and high-volume 'tastings'. The pandemic gave the industry a chance for a hard reset. Many have swapped their decor from a rustic country kitchen to modern farmhouse. Seated tastings in private tents, charming patios and flower-filled gardens have flourished. Many tasting rooms have turned away from buses, opting to serve small, more serious groups instead.

Recent years have seen several properties change hands, many of the region's founders selling to successful neighbours (such as Palmer Vineyards to Paumanok – see opposite), as well as moneyed developers and TV personalities. Serious financial investment has flowed into building renovations and vineyard expansion. Luxury touches nearly every wine region in America, so a strain of affluence has seeped into the North Fork.

To get a taste of innovation in a traditional country setting, head to Paumanok (*paumanok.com*) in Aquebogue, a 10-minute drive east from Riverhead. Paumanok pays homage to the Algonquian word for Long Island, which translates roughly to 'the island that pays tribute'. Local pioneers Ursula and Charles Massoud planted vines in 1983. As the second generation, Kareem Massoud has earned accolades as a forward-thinking winemaker, praised for showcasing the potential of Chenin Blanc on Long Island. In 2018, the family acquired Palmer Vineyards, where Massoud makes the wine, including one of the region's best Albariños.

Macari's Dos Aguas red

Paumanok offers several tasting experiences. At the top end, Kareem leads a VIP flight during which he shows his line of minimal-intervention wines and explains regional vintage variation. 'We had our best

Macari Vineyards' Bergen Road Bungalows, near Mattituck (see p76).

vintage in years in 2022,' he recently declared. Book the deck of the weathered farmhouse with a wine flight and vineyard views for a relaxing afternoon.

Leaving Paumanok, it's about 10 minutes to another family-owned winery, Macari Vineyards just outside Mattituck *(macariwines.com)*. Founded by Joseph Macari Sr in the mid-1960s, with the first vines planted in 1995, the Macari family's waterfront farm, with its sweeping views of the Long Island Sound, is one of few in New York State to pursue regenerative agriculture and biodynamic farming.

Third generation and winery director of operations Gabriella Macari brings a global perspective to the local market. With the help of winemaker Byron Elmendorf, Macari experiments with pét-nats and skin contact whites, atypical for Long Island's conservative-leaning approach to winemaking. She also developed new tasting experiences like the Bergen Road Bungalows; heated tents dotted among the vines. 'I created them during Covid and expected that they would last one season, but the demand has been incredible,' she says.

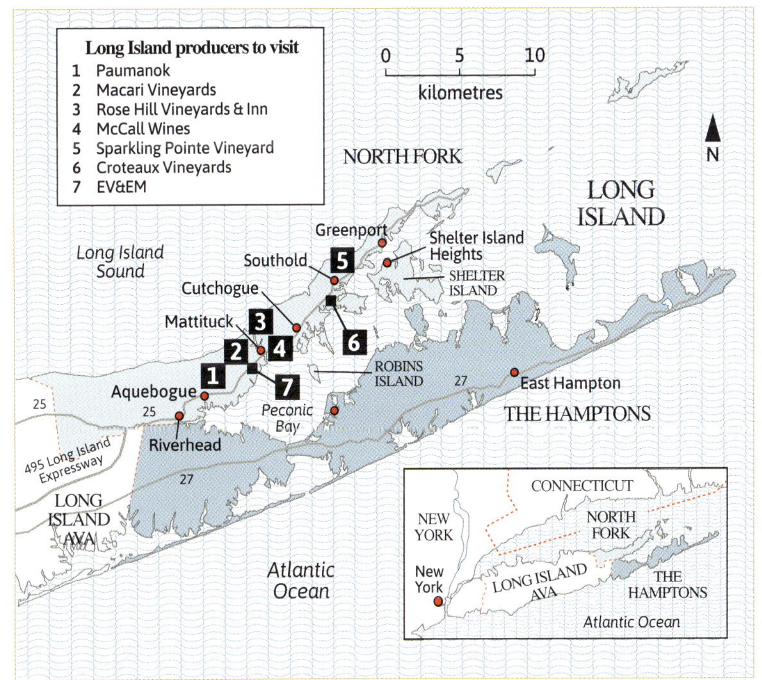

GETTING THERE

There are frequent domestic and international flights to NYC's three airports. Rent a car for the 160km drive to Greenport. Long Island Railroad (LIRR) trains depart New York Penn Station for Riverhead and Greenport, but you need a car to visit wineries.

Recently, the family opened the stylish Meadowlark North Fork, a tasting bar in Cutchogue serving limited and innovative labels such as Macari's Horses Cabernet Franc Pét-Nat 2021 and the flagship red blend Bergen Road 2020.

East of Macari by seven minutes, Rose Hill Vineyards & Inn *(rosehill-vineyards.com)* changed hands a few years ago. Shinn Estate Vineyards, founded by visionaries David Page and Barbara Shinn, was sold to financier Randy Frankel in 2017. He took over the charming farmhouse, four-room inn and 125-year-old barn on Oregon Road and gave it a top-to-bottom makeover, dropping the 'rustic' and adding the 'chic' to the interiors. Frankel kept long-time winemaker Patrick Caserta for his skill and consistency.

Bordeaux-style blends have long ruled the house style, and the same persists today. The Cabernet Franc and Wild Boar Doe – the latter composed of Merlot, Cabernet Sauvignon, Petit Verdot and Malbec – consistently prove that Long Island can make sensual, textural, structured reds with delineated flavours and sweeping freshness.

Though the climate of Long Island typically doesn't suit thin-skinned, disease-prone grapes, one winery has managed the delicate art of producing fine Pinot Noir. About five minutes southeast of Rose Hill sits McCall Wines *(mccallwines.com)* in Cutchogue. The tasting room occupies a polo horse stable, formerly a potato barn. The rustic aesthetic of exposed wood beams and worn saddles underlines the family's long-standing roots in wine, horsemanship, and agriculture. McCall hosts only small groups of wine connoisseurs.

One winery has gambled on traditional-method sparkling wine and won. Sparkling Pointe Vineyards & Winery *(sparklingpointe.com)* sits on a breezy spot about eight minutes east of McCall in Southold. Cynthia and Tom Rosicki produce estate-grown sparkling wine from 16ha planted with the classic Champagne trio of Pinot Noir, Meunier and Chardonnay.

'Luxury touches nearly every wine region in America, so a strain of affluence has seeped into the North Fork'

Despite world-class winemaking, luxury experiences – whether caviar pairings, private tents, or barrel tastings – were the exception not the rule here. Now, a touch of glamour has gilded the North and South Fork's antique barns and weathered farmhouses, attracting a new crop of wine tourists to Long Island's East End.

ABOVE Sparkling Pointe vineyard and tasting room building at Southold
RIGHT Rose Hill Vineyards & Inn, near Mattituck

Claudio's Waterfront, Greenport Harbor

The Menhaden

The Frisky Oyster

MY PERFECT DAY ON THE LONG ISLAND WINE TRAIL

MORNING

I always opt for Greenport as a base to explore the North Fork. This sleepy fishing village remains a quiet holdout compared to its South Fork counterpart, Montauk – like the swish black and white **The Menhaden** hotel *(themenhaden.com)*, where I wake to drink coffee in the general store downstairs. I contemplate a brisk bike ride on a hotel cruiser, but the urge for folded eggs on brioche with gruyère and sugar bacon from **Bruce & Son** *(bruceandsongreenport.com)* overrides exercise.

LUNCH & AFTERNOON

Blessed with a blue sky, I drive 10 minutes west to **Croteaux Vineyards** *(croteaux.com)*. Closed over winter, the beloved rosé-only brand serves several styles on a pea-gravel patio framed in sumptuous gardens. Très North-Fork-meets-Provence. I try to catch a few more wineries with refreshed looks courtesy of new management, including **EV&EM** *(evandemvineyards.com)* in Laurel Lake. A sleek, lounge-like tasting room replace the former era's tired, standing-room-only tasting bars. For lunch, I head to seasonal spot **Duryea's at Orient Point** *(duryeas.com)*. After a stuffed lobster roll on the Mediterranean-inspired terrace, a quick swim in the cold waters off the beach club is refreshing. On the way back, a break in East Marion enables a visit to **Lavender by the Bay** *(lavenderbythebay.com)*. The 9ha farm near the slate-blue sea has more than 80,000 plants for bouquets, essential oils and soaps. Visitors can walk the perfumed farm trails – just don't pick the flowers.

EVENING

Back in Greenport, I score a sunset table at **Claudio's Waterfront** *(claudios.com)* on the pier. A local IPA beer with a plate of mixed clams whet the appetite. Strolling back to **Little Creek Oyster Farm** *(littlecreekoysters.com)*, a humble shack in front of the iconic Bait & Tackle sign, I then stop by **The Frisky Oyster** *(thefriskyoyster.com)* and **Noah's** *(chefnoahs.com)* on Front Street – you can put your name on the short wait list, then grab a glass of wine at either bar. For a nightcap, **Brix & Rye** *(brixandrye.com)* on Main Street serves the best cocktails.

MENDOCINO, CALIFORNIA

In the centre of Mendocino County is the bucolic Anderson Valley, home to a string of renowned wineries, superb restaurants and effortlessly excellent places to stay, all within a short drive.

Now a destination in its own right, the Anderson Valley has developed into more than just a stopover for visitors on their way to the Pacific coast. And a good place to base yourself when exploring this northerly cool-climate wine region is the idyllic coastal town of Mendocino. It was founded in 1851 but fell into economic decline with the dissipation of the logging industry, and didn't really bounce back until the late 1950s, when it became a haven for artists. Since then, Mendocino's quaint charm has made it beloved by visitors near and far – a town frozen in time, thanks to much of it being on the National Register of Historic Places and many of its original buildings qualifying as California Historical Landmarks.

The Sea Rock Inn looks out over one of the cliffs that surround the picturesque town of Mendocino, on a remote stretch of coastline in northern California

Navarro Vineyards (p138) offers free tastings

The tasting patio at Pennyroyal Farm

About 50km inland, in the Anderson Valley, development was slower. It wasn't until the 1970s, more than a century after early settlers arrived here, that the first commercial wineries planted vineyards. Due to the cool weather, making it harder for thicker-skinned red wine grapes to fully ripen, the initial focus was on white varieties such as Riesling, Chardonnay and Gewürztraminer. This was still the case when Anderson Valley was officially designated an American Viticultural Area (AVA) in 1983.

While small amounts of Pinot Noir were planted in the 1970s, its popularity increased in the 1990s, and the potential for the grape in Anderson Valley was more widely recognised. Today, Mendocino County produces considerably more Pinot Noir than it did in the 1980s – some 1,200 tonnes were crushed in 1980 compared to more than 6,700 tonnes in 2021 – with much of that grown and made in the Anderson Valley.

Along the wine trail

In the centre of the valley, about 60km southeast of Mendocino along Highway 128, is Pennyroyal Farm (pennyroyalfarm.com) one of the region's newest estates. Proprietor and head winemaker Sarah Cahn Bennett – daughter of the founders of Navarro Vineyards (see p138) – practises regenerative and sustainable farming on her 40ha property, growing vegetables, herbs and fruit, including 10.5ha of Pinot Noir and Sauvignon Blanc. The Pennyroyal team hosts tastings by reservation on the outdoor patio from Thursday to Monday, and picnic tables are available to those wanting to stop for a glass of wine and something from the 'Farm Fare' menu.

'A favourite with visitors for almost five decades, Navarro is also one of the few wineries in the valley where tastings are still free'

Driving back up the highway toward Philo you'll arrive at the Drew Family Wines *(drewwines.com)* tasting room at The Madrones complex, where visitors can enjoy award-winning Pinot Noir and Chardonnay made from fruit grown primarily in the winery's cool, fog-shrouded, high-elevation Mendocino Ridge vineyards. The tasting room (where you're likely to find owners Molly or Jason Drew serving) is open by appointment on weekends but walk-ins are accommodated if space permits.

Toulouse Vineyards *(toulousevineyards.com)* is just a two-minute drive back up the highway. Offering one of the nicest views in the valley from the rear deck, overlooking the Navarro river and Hendy Woods State Park, it is the perfect place to stop on a sunny day for an afternoon tasting of aromatic whites and Pinots. The ambience here is relaxed and fun, just the way that

FACT FILE: MENDOCINO COUNTY

PLANTED AREA (2020) 7,070ha
MAIN GRAPES (BY AREA)
Red Pinot Noir, Cabernet Sauvignon, Zinfandel, Merlot, Syrah, Petite Sirah, Carignan, Grenache, Sangiovese, Barbera
White Chardonnay, Sauvignon Blanc, Gewürztraminer
AVAs
Anderson Valley, Cole Ranch, Covelo, Dos Rios, Eagle Peak, McDowell Valley, Mendocino, Mendocino County, Mendocino Ridge, North Coast, Potter Valley, Redwood Valley, Talmage Region, Ukiah Valley, Yorkville Highlands

SOURCE: MENDOCINO WINEGROWERS

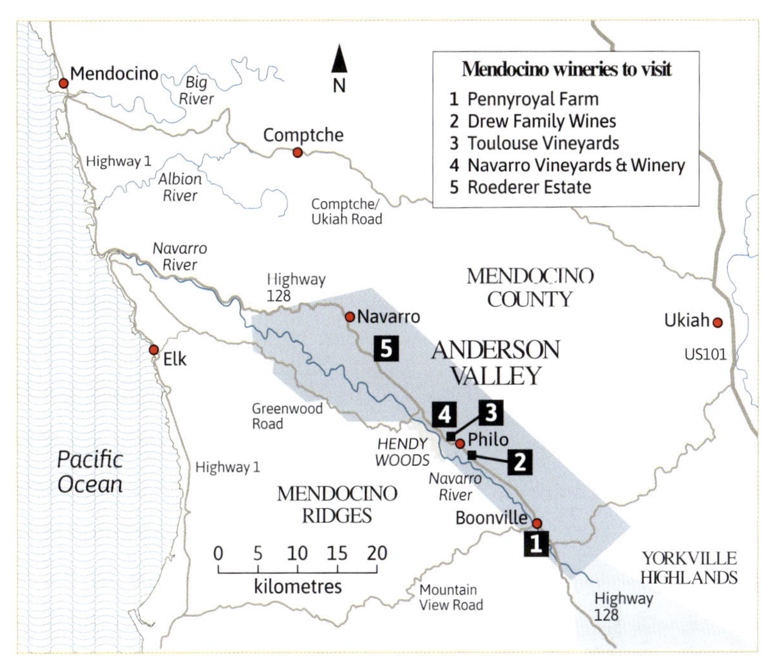

GETTING THERE

It's a 2.5- to 3-hour drive from San Francisco International Airport to the Anderson Valley. Or it's just over an hour's drive from the smaller Sonoma County Airport, which offers flights to/from US destinations including Los Angeles, San Diego, Portland, Seattle, Las Vegas and Phoenix.

MY PERFECT DAY ON THE MENDOCINO WINE TRAIL

LUNCH & AFTERNOON

Arrive in Boonville, in the centre of the Anderson Valley, and enjoy a farmstead cheese and charcuterie board with a wine tasting flight at **Pennyroyal Farm** (see p136). After lunch, head back up Highway 128 to **Navarro Vineyards** (p138) for a wine tasting on the patio. From there, take a 40-minute drive towards the coast to check into your room at **The Harbor House Inn** (p138) in Elk, south of Mendocino, before enjoying a walk down to the inn's private cove and beach to search for abalone shells or watch for dolphins, sea lions and birds at play.

EVENING

Take your seat at The Harbor House restaurant, which offers views of the chef's gardens and the coastline. This isn't just a meal, though – it's an interactive experience, as your server and sommelier describe each dish and wine and the origin of the ingredients.

End the day with a book by the fire in your room, or bundled up on your deck, stargazing and listening to the ocean.

NEXT MORNING

After breakfast, visit the charming town of Mendocino for a scenic walk along the remote, rugged headlands, beginning at the visitor centre. Breathe in the crisp air mixed with the fresh scent of saltwater, bay and redwoods and let the hypnotic sounds of crashing waves lull you into a zen-like state. After your walk, stop at the **Goodlife Cafe & Bakery** *(goodlifecafemendo.com)* in the centre of town for a freshly made sandwich, a hearty salad or a bowl of soup.

owners Vern and Maxine Boltz like it. You'll find one of the winery's slogans proudly displayed on a sign outside: 'Too tense? Toulouse'.

A further three-minute drive along Highway 128 brings you to Navarro Vineyards & Winery *(navarrowine.com)*, set against a gorgeous backdrop of rolling hills and vineyards. Founded in the 1970s by Ted Bennett and Deborah Cahn, Navarro was one of the first commercial wineries in the Anderson Valley, and is best known for its aromatic, Alsace-style white wines and friendly, casual atmosphere. A favourite with visitors for almost five decades, Navarro is also one of the few wineries in the valley where tastings are still free.

Roederer Estate *(roedererestate.com)* is just up the road, about 45km from Mendocino. While the roots of this property are tied to the venerable Champagne house that created Cristal, the estate itself is humble, choosing to highlight the natural beauty and relaxed style of the Anderson Valley. Seated tastings are offered indoors or outdoors and feature flights of the winery's estate-bottled traditional-method sparkling wines, with or without food prepared by the winery's chef. At weekends, visitors can also enjoy a picnic with bottle service.

YOUR MENDOCINO WINE COUNTRY ADDRESS BOOK

ACCOMMODATION

Little River Inn

On Highway 1, just 5km south of the town of Mendocino, this fifth-generation, family-run oceanfront hotel and resort gives guests an expansive view of the Pacific, comfortable rooms equipped with jacuzzi tubs and fireplaces, and an on-site restaurant and golf course. There are also pet-friendly rooms. **littleriverinn.com**

The Boonville Hotel

A cosy, understated but upscale rural roadhouse, this is an institution in the Anderson Valley. Originally built in the late 1800s, it has 17 accommodation options, including a few 'casitas' or cabins, offering additional privacy and space. The concept throughout is 'simplicity by design', rejecting clutter for quality. **boonvillehotel.com**

The Harbor House Inn

Perched on a cliff overlooking the Pacific, The Harbor House Inn features six luxurious rooms and five private cottages, a two-star Michelin restaurant (see right), expansive gardens and access to a private cove and beach. Built in the early 1900s and renovated in 2018, the property exudes effortless and comfortable elegance. **theharborhouseinn.com**

RESTAURANTS

The Harbor House Inn Restaurant

Run by chef Matthew Kammerer (among the first to earn a Michelin Green Star), this offers an immersive dining experience. Ingredients are sourced from within a small area, focusing on regenerative farming methods. For dinner, the Full Experience tasting menu is paired with mainly local and often rare wines. **theharborhouseinn.com**

Wickson

Opened in late 2020, this is a partnership between chefs Alexa Newman and Rodney Workman and the owners of The Madrones complex where it is located. Wickson is an homage to the Wickson apple variety, historically grown in the Anderson Valley before grapes became the main crop. The focus is on wood-fired dishes and pizzas, and on working with the farmers, foragers, fishermen, brewers and vintners of Mendocino and Sonoma counties. **wicksonrestaurant.com**

SHOPS

Farmhouse Mercantile, Boonville

Featuring items made or designed by local artists, this has the look of a country store and the feel of a workshop. It features simple but expertly crafted items such as handmade plates and bowls, leather bags, kitchen linens and carved utensils made (as far as possible) using natural resources and repurposed materials. **farmhouse128.com**

Mendocino Art Center

Founded in 1959 by artists who spurred the cultural revival of the town, this is a gallery showcase of paintings, photography, sculptures, ceramics and other mixed media creations by local talent. The centre runs regular exhibitions, classes and workshops, and sponsors several artists in residence throughout the year. **mendocinoartcenter.org**

Little River Inn hotel and resort,
ocean views at sunset

US WINE ROADS LESS TRAVELLED

Taste the freewheeling spirit of American winemaking in these five 'other' US states...

America's big-name wine states – California, Washington, Oregon, New York and Virginia – are well known to wine lovers. But smaller (though no less quality-driven) appellations in **Idaho**, **Colorado**, **Texas**, **North Carolina** and **Michigan** are among other exciting US wine destinations worth exploring.

While vastly different, these less-famous wine states share some important universal truths. For a start, you won't find roads clogged with chauffeured limos, and your tasting fees are more likely to be $5 rather than $50 per person. The wineries you'll visit are almost always intimate, family-run operations, and more often than not the person whose name is on the bottle will be the one pouring you a sample.

Most importantly, the wines will be anything but predictable – expect to find intriguing riffs on less-famous grape varieties such Malvasia, Lemberger and Dornfelder.

A true taste of America's entrepreneurial soul is yours to enjoy when you take these wine roads less travelled.

IDAHO

With its intriguing mix of volcanic and glacial soils, Idaho was once a promising wine-growing region. Indeed, some of the first vines planted in the Pacific Northwest were established in Lewiston, Idaho in the 1860s. Today, the state is home to 65 vineyards, 69 wineries and three AVAs: Eagle Foothills (an AVA within Snake River Valley with 32ha under vine), Lewis-Clark Valley and Snake River Valley.

There are more cows than people here, and the state's defining quality is its dramatic geologic history, which inspires the vivid-sounding nomenclature of destinations such as Hell's Half Acre, Sawtooth Mountains and Lava Hot Springs.

Idaho is a basin where glaciers melted and stayed, leaving a mix of sediment and soil. Several young volcanoes also influence the wines, says Ste. Chapelle Winery winemaker Meredith Smith: 'We have cinder pits from volcanoes in our vineyards and they impart special flavours, such as dark red fruit notes, spice and cigar box.'

Its arid climate, long sunny days and cool nights, together with high elevation (upwards of 900m) are distinctive qualities. With 728ha under vine, the largest AVA, Snake River Valley (which also overlaps into Oregon state to the west) is spread across the cities of Caldwell, Boise and Garden City – all within a short drive from one another. 'Snake River Valley is on the same latitude as Rioja in Spain and the northern Rhône,' explains Earl Sullivan, winemaker for Telaya Wine Co (*telayawine.com*), 'so Tempranillo, Syrah and Viognier do really well.'

With just 39ha under vine, Lewis-Clark Valley AVA in northern Idaho (also overlapping into Washington state) sits at the base of the Bitterroot Mountains. Vines were planted here as early as 1872, and one might still stumble across abandoned vineyards. Lewis-Clark Valley is part of Idaho's 'banana belt', where more temperate conditions favour varieties such as Cabernet Sauvignon and other late-ripening Bordeaux grapes.

LEFT Ste. Chapelle Winery at Sunnyslope, western Idaho and barrels at Clearwater Canyon, which runs preview tastings for its club members

IDAHO: STAY & EAT

Inn at 500 Capitol (*innat500.com*) is a contemporary property in the centre of Boise, with luxurious touches such as fireplaces and private balconies, and three top-floor penthouse suites. **Richard's Restaurant & Bar** (*richardsboise.com*) is located in the hotel and serves sophisticated Italian fare along with a wine list that features several Idaho wines. Book early as it's a local favourite.

Located in Boise's hip Linen District, **The Modern Hotel and Bar** (*themodernhotel.com*) is a stylish, reimagined ex-Travelodge, owned and operated by a Basque family. (Boise has the largest concentration of Basques outside Spain.)

The Grove Hotel (*grovehotelboise.com*) is one of Boise's most luxurious properties, located in the heart of downtown and within walking distance of almost everything. It has a spa, hot tub, pool and a beautifully upscale restaurant, **Trillium**. In Caldwell, on the Sunnyslope Wine Trail, visit **Grit** (*grit2c.co*) for inspired comfort food – the slow-fried chicken is a must-try.

The most logical base for touring Idaho wineries is the capital city of Boise and the nearby Sunnyslope Wine Trail (about 30 minutes' drive away). Here, you can rent a bike and visit several wineries via the 40km Boise River Greenbelt. Stop for a private tour and tasting at **Telaya Wine Co** – its Turas 2018, a blend of Syrah, Petit Verdot, Merlot, Tempranillo and Sangiovese, won Best Red Wine at the 2020 Idaho Wine Competition.

In Sunnyslope, look for the richly textured Panoramic Shoshone Falls Malbec 2018 from **Ste. Chapelle Winery** (*stechapelle.com*), one of Idaho's oldest producers and a standard-bearer for the Snake River Valley terroir. It also runs concerts and has yurts for overnight stays. Next door, **Sawtooth Winery** (*sawtoothwinery.com*) also hosts concerts and themed dinners, or just visit to taste its Classic Fly Riesling or Sparkling Brut.

At nearby **Williamson Orchards & Vineyards** (*willorch.com*), you'll have the chance to pick cherries and sip its wonderfully delicate Albariño 2019, a Best in Show winner at the 2020 Idaho Wine Competition.

COLORADO

High in elevation and anchored by the visually stunning Grand Mesa, the largest flat-top mountain in the world, western Colorado wine country is a study in superlatives. The state has two official AVAs, Grand Valley and West Elks, which range in elevation from 1,200m-2,100m, making them among the highest vineyards in the world. Indeed, Colorado's high desert has captivating scenery but it tends to be overshadowed by the deluxe appeal of the state's many ski towns. Here, close to the Utah border, you'll find fewer tourists, more locals, the world's second-largest concentration of geologic arches (Rattlesnake Canyon), and plenty of award-winning wine.

With a truncated growing season, unpredictable weather patterns and dramatic diurnal temperature changes (upwards of 10°C), many vintners describe winemaking here as a challenging adventure. 'We never know what Mother Nature is going to throw at us. You can't set a clock by nature's whim the way you can in California wine country,' explains Kevin Webber, co-founder of **Carboy Winery** *(carboywinery.com)*, one of Colorado's largest wineries.

The Grand Valley's arid, high-elevation setting has soils that tend to be more alkaline than those of Napa, yielding wines that taste more Old World than New World. Syrah, Viognier and other Rhône varieties fare well in Grand Valley, as do Bordeaux grapes, especially Cabernet Franc.

Not far from Grand Valley, snug in the embrace of the Rocky Mountains, the West Elks AVA is home to true mountain terroir, with grapes cultivated at altitudes as high as 1,950m. The result is varieties such as Riesling, Pinot Gris and Pinot Noir, which can tolerate a cooler growing season with intense UVA sunlight.

The Grand Valley region is home to a mix of the old west and the new west, blended with the region's historic mining culture. The towns of Grand Junction and Palisade are 12 minutes apart and either one makes a good base for wine touring – several wineries have tasting rooms in both towns. In Colorado, getting outside is almost a requirement, so tour the tasting rooms in your hiking boots, or rent bikes and visit on two wheels. Grand Valley is one of the state's top mountain-biking destinations; it's also ideal for canyoneering, white-water rafting and lengthy hikes. West Elks, which is just over an hour's drive from Grand Junction, has a smaller wine trail featuring 10 wineries, making it an easy day trip.

Carboy is betting on sparkling wine and cultivating more cold-hearty hybrids. At its recently opened tasting room in Palisade, start with the Native Fizz Rosé, a co-fermented blend of North American hybrid grapes Verona, Aromella and Vignoles. **Red Fox Cellars** *(redfoxcellars.com)* is another fine choice to quench your thirst in Palisade. Sip a wine cocktail, sample one of its eight on-tap craft ciders, or try its Nebbiolo 2017, a Governor's Cup double gold medal winner in 2019.

Across the river is the **Colterris Winery** at the Overlook (colterris.com) tasting room, panoramic views of the Colorado river and rows of lavender and roses forming a beautiful backdrop – and the wine measures up as well. Order a charcuterie sampler and a bottle of the Petit Verdot 2017 – another Governor's Cup double gold winner. Around the corner, **Maison la Belle Vie Winery** *(maisonlabellevie.com)* makes a sumptuous Vin de Peche, a Muscat fortified with peaches, from a family recipe that dates back to the late-1800s. Ask about its wine-paired dinners as well.

In West Elks, **The Storm Cellar** winery *(stormcellarwine.com)*, founded by sommeliers Jayme Henderson and Steve Steese, focuses on aromatic whites and rosés. Tastings are intimate affairs and private tastings are an option too. Top tastes include the 2019 RRV, a mouthwatering blend of Roussanne, Riesling and Viognier.

COLORADO: STAY & EAT

In Grand Junction, locals love **Bin 707 Foodbar** *(bin707.com)*, where chef/owner Josh Niernberg (a 2020 James Beard Best Chef Mountain Region semi-finalist), serves up fresh, inspired Colorado cuisine from the best local provisions he can source. **Hotel Maverick** *(thehotelmaverick.com)* is a new, nicely appointed upscale hotel located on the campus of Colorado Mesa University. It offers a rooftop restaurant, and easy access to wine trails.

In the small, unassuming town of Palisade, **Pêche** *(pecherestaurantcolorado.com)* is an exquisite culinary discovery, serving simple and fresh ingredients, artfully prepared. Try the Thai fried chicken paired with Storm Cellar Riesling and finish the meal with a portion of its marvellous rhubarb cheesecake.

For an overnight stay in Palisade, **Spoke & Vine** *(spokeandvinemotel.com)* is a brilliantly renovated motel that offers a hip, friendly, no-fuss vibe. The owners take pride in the good coffee, plush beds and being pet-friendly, so leave any pretensions at the door.

'We never know what Mother Nature is going to throw at us'.

Kevin Webber, **Carboy Winery**

Mount Garfield and vineyards at Palisade in Grand Valley, western Colorado.

TEXAS

Texas is the second largest US state. For perspective, it's about 20% larger than France, and has roughly 200 wineries and eight official AVAs (Bell Mountain, Escondido Valley, Fredericksburg, Mesilla Valley, Texas Davis Mountains, Texas High Plains, Texas Hill Country and Texoma). Despite its impressive size, winemaking is chiefly concentrated in two distinct areas: Texas Hill Country and High Plains. Most of the state's grapes are cultivated in High Plains, while most of the tourism and consumption takes place in Texas Hill Country. Spend a minute in the High Plains AVA and you'll appreciate why – dominated by semi-arid, windy conditions, it's a high-elevation, vastly flat region of roughly 3.2m hectares in west Texas. Andrew Sides, winemaker for Lost Draw Cellars, describes it as 'ideal for winemaking but not for the faint of heart'. The area has rich, sandy loam soils – primarily ancient seabed, underneath which is a deep limestone bed that imparts an intriguing minerality to the wines.

The High Plains may be the state's primary growing region, but it's the scenic charms of Texas Hill Country that draw the crowds. The climate here is more moderate, and the soils are more on a granite uplift. Concentrated around the town of Fredericksburg (which is equidistant from San Antonio and Austin), Hill Country enjoys a lingering Germanic influence as many Germans settled here in the mid-1800s. It is also the home of former president Lyndon Johnson and known for its profusion of bluebonnet wildflowers in the spring. Warmer-climate varieties such as Viognier and Tempranillo thrive in both High Plains and Hill Country. You'll also find Sangiovese, Roussanne, Mourvèdre and some Tannat.

At **William Chris Vineyards** *(williamchriswines.com)*, 32km east of Fredericksburg, a glass-walled tasting room looks out over the vineyards. Order a picnic lunch if you want to enjoy wines al fresco, or consider a seat inside for the Library Tasting experience that features an in-depth look at the range and breadth of its terroirs and vintages.

The Signature Series of wines at **Pedernales Cellars** *(pedernalescellars.com)*, back towards Fredericksburg, features delicious single-barrel expressions, best enjoyed in the tasting room with views over the Pedernales river valley. **Becker Vineyards** *(beckervineyards.com)* has a long history of winemaking in Hill Country (former first lady Ladybird Johnson was a fan of its Chardonnay). Book a tour and reserve library tasting paired with cheeses for the best experience, and if you visit in the spring you'll enjoy the full bloom of its fields of wildflowers and lavender plants.

Another worthy stop is at **Bingham Family Vineyards** *(binghamfamilyvineyards.com)*. It scooped five awards at the 2021 San Francisco Chronicle Wine Competition – including gold for its Dugout 2018 Bordeaux blend.

The charming courtyard at **Lost Draw Cellars** *(lostdrawcellars.com)* in Fredericksburg is a big draw for live music at weekends, and a glass or two of its Reserve Roussanne (2018). Book a wine and charcuterie tasting if you want some light bites.

BELOW Co-founder and winemaker David Kuhlken of Pedernales Cellars.

'It's the scenic charms of Texas Hill Country that attract the crowds'

LEFT The tasting room at William Chris Vineyards.

BELOW Visitors enjoy a stop on the Texas Hill Country wine trail; Lost Draw Cellars, THP Rosé blend.

TEXAS: STAY & EAT

The universal favourite for overnight stays is **Hoffman Haus** *(hoffmanhaus.com)*, a luxury bed-and-breakfast well situated in the heart of Hill Country. **Outlot 201 Guest Houses** *(@outlot201GH)*, an 8km drive from Fredericksburg's historic Main Street, offers three guest houses designed in keeping with the area's traditional 'Sunday Haus' style homes. Each one has a pantry stocked with homemade banana bread, fresh fruit and drinks. For a boutique hotel experience, the Trueheart Hotel (thetruehearthotel.com) offers 13 rooms furnished in a colourful and playful style. Or for something quirkier, consider the adults-only 1940s aviation-centric **Hangar Hotel** *(hangarhotel.com)*.

You'll have your pick of places to eat and drink in Fredericksburg, but a dish of goose and truffle ravioli with mixed farm greens, wine reduction sauce and pecans at **Otto's German Bistro** *(ottosfbg.com)* tops the list for a dose of the region's Hessian history. **Cabernet Grill** *(cabernetgrill.com)* is popular with locals, and has one of the largest selections of Texas wines in the state – a great place to taste options not on the Hill Country tasting trail.

NORTH CAROLINA

The geography in this long and narrow state ranges from mountains to coastline; in between the two is where the finest winemaking happens – the North Carolina Piedmont. The entire state has more than 200 wineries and six AVAs (Appalachian High Country, Crest of the Blue Ridge, Haw River Valley, Swan Creek, Upper Hiwassee Highlands and Yadkin Valley) spread across a wide distance; it is a daunting state to explore and grape expressions vary widely. In fact, explains Louis Jeroslow, owner of Elkin Creek Vineyard in Yadkin Valley: 'We haven't found many grape varieties that don't do well here. It's an exciting time where people are planting everything they can get their hands on. Right now, variety diversity is the defining feature in North Carolina.'

For the purposes of touring, and to ensure access to a nice density of wineries, Yadkin Valley and Swan Creek offer the greatest opportunity. This slice of North Carolina is the envy of the state with its serene, smooth-edged mountains. Located northeast of Asheville, North Carolina's groovy, beer-loving town, Yadkin Valley is the oldest, most established AVA in North Carolina (circa 2003). There are 48 wineries in the region and vines here grow in the type of sandy clay found in Tuscany, while the temperature and humidity averages are almost exactly like those in Bordeaux. 'These vines don't know they are in North Carolina,' says Jeroslow. 'Their roots are in Italy and the fruit and leaves are in France.'

During your exploration of Yadkin, you can take an interesting detour into the Swan Creek AVA, a sub-appellation of Yadkin Valley. These wines tend to have a distinctive mineral note – thanks to the area's proximity to the Brushy Mountains, an isolated spur of the Blue Ridge Mountains with unique mineral deposits.

NORTH CAROLINA: STAY & EAT

The Rockford Inn *(rockfordbedandbreakfast.com)* is an historic antebellum home that dates back to 1848. Located in the village of Rockford, it offers a quiet escape, with pretty gardens and easy access to the Yadkin Valley wineries. Nearby, the Wine Lodge at **Stony Knoll Vineyards** *(stonyknollvineyards.com)* is a carefully restored 1890s homestead that features two winery houses overlooking the vines at Round Peak Vineyards.

For a more rustic vibe, consider a stay at **Klondike Cabins** *(theklondikecabins.com)* at Grassy Creek Vineyard (grassycreekvineyard.com), 24km west; they were formerly the hunting cabins for the Hanes family of Hanesbrands textile fame.

In Pilot Mountain, **End Posts Restaurant at JOLO Winery & Vineyards** is a wonderful option for a lunch of shared plates and tapas. **Harvest Grill at Shelton Vineyards** is a local favourite, while on Sundays, **Elkin Creek Vineyard** serves up brick oven gourmet pizzas, but plan ahead and make a reservation – it's very popular.

LEFT One of the Klondike Cabins at Grassy Creek; the tasting room at Pilot Mountain Vineyards.
BELOW JOLO Winery & Vineyards lies close to the domed Pilot Mountain, Yadkin Valley; vines extend over 4ha at Stony Knoll Vineyards.

Kick off your Yadkin Valley wine trail with stops at the new Pilot Mountain Winery & Vineyards *(pilotmtnvineyards.com)* and JOLO Winery & Vineyards *(jolovineyards.com)*. Both wineries enjoy amazing views of the distinctive Pilot Mountain – a quartzite dome that's unlike any other mountain in North Carolina and the defining geographical feature in Yadkin Valley.

Further north, Round Peak Vineyards *(roundpeak.com)* is popular for both its Nebbiolo and its 18-hole disc golf course that wanders through the vineyards; it has two cabins for rent, and a brewery. Visitors to Shelton Vineyards *(sheltonvineyards.com)* enjoy the local speciality sonker with their wine at the winery's Harvest Grill restaurant– it's a cobbler-style dessert handed down through generations in the area.

'We haven't found many grape varieties that don't do well here'.
Louis Jeroslow, **Elkin Creek Vineyard**

At Elkin Creek Vineyard *(elkincreekvineyard.com)* they'll pour the flagship wine, the Bordeaux blend Rossa (2017), but Jeroslow admits it's the Dornfelder that sells like hotcakes. 'We make it in the traditional German table-wine style, just a touch off dry to accent the herbal character; visitors love it,' he says. Llama-trekking is on offer at Divine Llama Vineyards *(divinellamavineyards.com)*, where you can sip the well-regarded Reserve Merlot before taking a 3km llama trek – a perfect way to explore the mountainous terrain.

MICHIGAN

Michigan winters are notoriously brutal and yet its winemaking industry thrives, owing much of its success to a particular microclimate created by ancient glacial activity. While there are five AVAs in the state (Fennville, Lake Michigan Shore, Leelanau Peninsula, Old Mission Peninsula and Tip of the Mitt), the two most prominent (Leelanau, Old Mission Peninsula) got their start more than 10,000 years ago as powerful glaciers formed the Great Lakes and created the peninsulas.

Surrounded by the waters of Lake Michigan, the Leelenau Peninsula and Old Mission Peninsula AVAs enjoy the moderating influence of the lake effect on winter's bitter temperatures, extending the harvest season and ultimately benefiting the aromatic, acid-driven varieties that do best here. Traverse City is the central point between both AVAs, which together comprise the Traverse Wine Coast, an easily navigable tasting trail comprising 40 wineries, making it the largest collection of winemakers in the Midwest.

The glaciers may have done the landscaping, but most locals agree the Michigan wine industry really owes its reputation to the late Ed O'Keefe, the founder of Chateau Grand Traverse winery *(cgtwines.com)*. According to Mike Kent, public relations manager for Traverse City Tourism, in the 1970s, O'Keefe 'had the crazy idea that because of the moderate microclimate on these peninsulas, you could pursue winemaking. He was right, and in just 30 years we've seen tremendous growth and our wines have won many prestigious awards'.

Today, the Traverse Wine Coast and its two peninsulas are home to 60% of Michigan's total wine production. The focus in this area is on cool-climate aromatic reds and whites such as Riesling and Pinot Gris, as well as rosé made from Pinot Noir and Cabernet Franc – and the occasional Lemberger. Incredible views of Lake Michigan framed by picturesque hillsides topped with quaint lighthouses can be found throughout the region.

Vines at the Ciccone Vineyard & Winery on the Leelanau Peninula

Any visit to this area must include a stop at O'Keefe's pioneering winery. It's a great place to watch the sunset while enjoying a flight of wines paired with seasonal menu offerings. It also has accommodation – The Inn at Chateau Grand Traverse – if you want to make a night of it. Left Foot Charley *(leftfootcharley.com)* in Traverse City has a century-old historic root cellar where it now ages its wines. Try the Sparkling Island View Vineyard Pinot Blanc 2018, crafted from Michigan's oldest Pinot Blanc planting (dating back to 1995).

On the Leelanau Peninsula, Ciccone Vineyard & Winery *(cicconevineyard.com)* is well-known for red expressions, especially the Cabernet Franc and Lee La Tage Bordeaux blend, both medal-winning wines. You may recognise the name – the owner and founding winemaker is pop star Madonna's father. Black Star Farms *(blackstarfarms.com)*, recognised for its quality Riesling, has wineries on both peninsulas, but the Suttons Bay location (5km away) delivers an all-in-one winery experience with a luxury inn (Inn at Blackstar Farms), restaurant (Hearth & Vine Café), and access to several hiking trails spread over 64ha. Just a 10-minute drive away, 45 North Vineyard and Winery *(fortyfivenorth.com)*, so named for the 45th latitude line that runs right through its winery, invites visitors to wander a lovely 5km trail that winds throughout the vineyards; or take a seat by the fire in the tasting barn, featuring handcrafted posts and beams, and indulge in the lemon cream notes of its extremely popular Unwooded Chardonnay.

MICHIGAN: STAY & EAT

Traverse City and its wine coast are well known as havens for delicious farm-to-table cuisine. If you love a food truck, visit **The Little Fleet** *(thelittlefleet.com)*, a permanent selection of food trucks serving up everything from local brews and wine to burgers and barbecue. More upscale, consider the waterfront dining option at the **Boathouse** *(boathouseonwestbay.com)*, on Old Mission Peninsula – pair the local speciality, smoked whitefish pâté, with a crisp unoaked Chardonnay.

Village Cheese Shanty *(villagecheeseshanty.com)* in Fishtown, Leland, features more than 60 types of cheese and local cherry preparations; people have been known to drive long distances to eat one of its epic sandwiches.

For an immersive wine country stay, book at **Chateau Chantal** *(chateauchantal.com)* on Old Mission Peninsula, which offers cooking classes and wine dinners, and rooms with a view over Grand Traverse Bay. **Park Place Hotel** *(park-place-hotel.com)* is one of the oldest hotels in town and a landmark in Traverse City, while the sprawling **Grand Traverse Resort and Spa** *(grandtraverseresort.com)*, 13km from Traverse City, is owned by Ottawa and Chippewa Indians, offering everything from gambling to golf.

CLOCKWISE FROM TOP At Chateau Grand Traverse, vines overlooking Grand Traverse Bay; the Boathouse restaurant on Old Mission Peninsula; Left Foot Charley, on the outskirts of Traverse City.

WALLA WALLA VALLEY

Mainly in Washington State but crossing into Oregon, too, the Walla Walla Valley AVA is an exciting place to explore, with a wide choice of tasting rooms to visit and places to eat.

Vineyards in the Walla Walla Valley AVA, which is home to about 120 producers

'The Walla Walla Valley's ascent from farming to renowned wine region has happened largely over the past 30 years'

Situated between the Blue Mountains in the southeast and the low, rolling hills of the Palouse to the north, the Walla Walla Valley (which straddles two states, Washington and Oregon) has an interesting topography. While it feels in a low position compared with the nearby mountains and hills, the Walla Walla Valley AVA – itself a sub-region of the larger Columbia Valley AVA – is actually at 300m altitude.

It also has an ideal climate for wine-growing. Four hours' drive east are the dry, sunny plains of Idaho, while the same distance west is coastal Seattle, boasting more rainy days a year than almost anywhere else on the US West Coast.

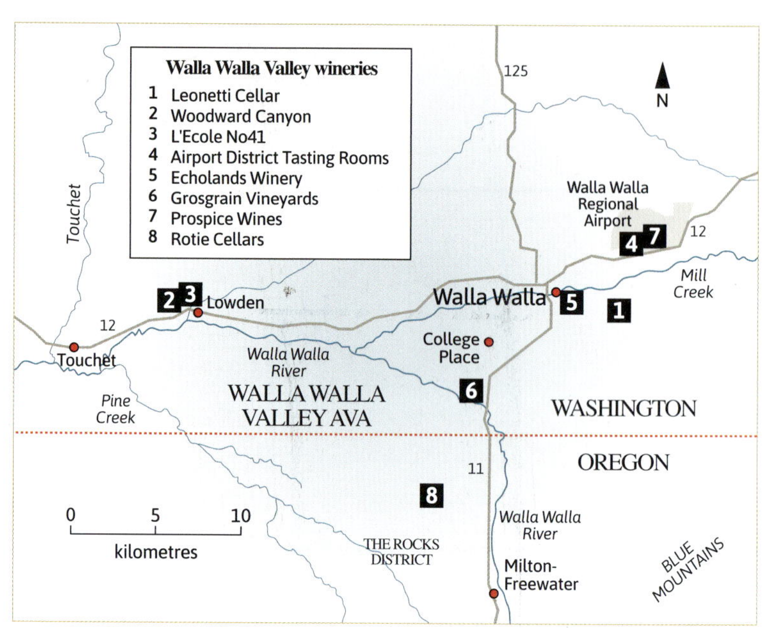

GETTING THERE

Fly to Walla Walla Regional Airport from Seattle (two flights a day; flight time 1 hour). By car, it's about 4 hours from both Seattle and Portland.

Visitors enjoy a tasting at Rotie Cellars' Rotie Rocks Estate

WALLA WALLA VALLEY: TOP TASTING ROOMS TO VISIT

AIRPORT DISTRICT TASTING ROOMS
Various locations, Walla Walla, WA 99362

On land that was once an airfield during World War II, the Airport District is now home not only to the Walla Walla Regional Airport but also a collection of wineries, breweries and distilleries operating out of renovated hangars, mess halls and garages. It also houses the Port of Walla Walla Regional Airport Wine Incubator Program, which helps start-up wine brands launch their businesses affordably. Current micro-winery incubator tasting rooms include: The Ink by Eternal Wines & Drink Washington State *(eternalwine.com; drinkwashington.wine)*, the 100% rosé-focused Smak Wines *(smakwines.com)*, Golden Ridge Cellars *(goldenridgecellars.com)*, Hoquetus Wine Co *(hoquetuswine.com)* and Itä Wines *(itawinery.com)*. Opening times vary: check each trader's website, and see **wallawalla.org**.

ECHOLANDS WINERY
7 West Alder, Walla Walla, WA 99362

Doug Frost, a Master of Wine and Master Sommelier, started this project in 2018 with friend Brad Bergman and brought on Taylor Oswald as winemaker. Their vineyards and winery are actually across the state border in Oregon, on the southern border of the Walla Walla Valley AVA, but this venue in downtown Walla Walla is their tasting room. Echolands sources fruit from sites within the AVA that produce ageworthy wines, such as Les Collines, Blue Mountain and Seven Hills. Varieties include Syrah, Grenache, Cabernet Franc, Merlot and Petit Verdot. Production is very limited, with only a few wines to try in the tasting room and some available only to wine club members. The tasting experience is educational yet relaxed. Open Thursday-Monday, 11am-5pm. **echolandswinery.com**

GROSGRAIN VINEYARDS
2158 Half Acre Lane, Walla Walla, WA 99362

This modern, artistic winery and tasting room (pictured, p154) has a southwestern bohemian vibe. Co-owner and fashion designer Kelly Austin's keen eye is evident in everything from the winery's furnishings and art to the colourful, textural and eclectic wine labels. The wines themselves are just as individual: don't expect big toasty reds or your usual Syrah and Bordeaux varieties. At Grosgrain, visitors will find a diverse collection of site-specific wines and rare offerings such as a crisp Albariño or a Lemberger Pét-Nat. Tasting appointments recommended, but walk-ins accommodated if available. Open Thursday-Monday, 10am-5pm. **grosgrainvineyards.com**

PROSPICE WINES
145 E Curtis Avenue, Walla Walla, WA 99362

Prospice ('look forward' in Latin) was founded in 2017 by Matt Reilly and Jay Krutulis from a shared desire to produce expressive wines from quality Washington fruit. The boutique winery has already found acclaim, particularly for its Resurgent Vineyard Syrah from an

elevated site near The Rocks District. To date, the winery's focus has been on Rhône varieties, including working with a local viticulturist and grower to find the right sites for Grenache in Walla Walla Valley. It also makes several wines from Bordeaux blends from top vineyards in the AVA. Reservations are required for the 60- to 75-minute tasting experiences, which include six wines. Open Friday-Sunday, 11am-6pm (5pm on Sunday). **prospice.wine**

ROTIE CELLARS
84328 Trumbull Lane, Milton-Freewater, OR 97862

Founded in 2006, Rotie Cellars – named after owner-winemaker (and former geologist) Sean Boyd's favourite northern Rhône region – was one of the first wineries in The Rocks District of Milton-Freewater to plant Rhône varieties. This sub-AVA, on the Oregon side of the border, is the only one in the US defined by its soil type: well-draining, gravelly, rocky soils that serve as heat conductors for the vines and contribute to lower yields and intense, floral wines. Boyd aims to make 'traditional Rhône blends with Washington State fruit' (although those with The Rocks designation are from Oregon). They all have moderate alcohols and see very little or no new oak. Tastings at the Rotie Rocks Estate tasting room. Open daily, sessions (by appointment only) at 11am, 12.30pm, 2pm and 3.30pm. **rotiecellars.com**

Grosgrain Vineyards
(see p153)

The Walla Walla Valley enjoys the middle ground: moderate rainfall, sunny days and slightly cooler temperatures than Washington's more westerly Yakima Valley AVA. But it's not that simple, as within the Walla Walla Valley there are clear variations. Its eastern border, which abuts the Blue Mountains, is colder than the west and gets three times more rain. This makes the valley's wines difficult to define in terms of regional typicity. Consequently there is a focus here on individual vineyards and their distinct terroirs.

To date, the Walla Walla Valley only has one official sub-AVA – The Rocks District of Milton-Freewater, on the Oregon side of the valley – but there are moves to define more.

The Walla Walla Valley's ascent from farming to renowned wine region has happened largely over the past 30 years. But three founding Washington wineries were making a name for themselves well before that: Leonetti Cellar (founded 1977), Woodward Canyon (1981) and L'Ecole No41 (1983). The region became an AVA in 1984, and between then and 2010, more than 100 new wineries joined those three pioneers. Today, there are about 120 producers in the Walla Walla Valley.

DOWNTOWN WALLA WALLA: WHERE TO EAT- AND MORE

AK'S MERCADO
21 E Main Street, Walla Walla, WA 99362

While there are many great spots for tacos in Walla Walla, AK's Mercado, which opened last year, is known for its large homemade corn tortillas as well as gluttonous menu items such as voodoo fries with pulled pork and jalapeños. Locals can't get enough of the braised short rib, smoked brisket, grilled fish, mole and braised pork tacos topped with pickled onions or veggies, cotija cheese and chipotle or serrano aioli. Open daily (except Wednesdays), 11am-10pm. **andraeskitchen.com**

KINGLET
55 W Cherry Street, Walla Walla, WA 99362

Opened earlier this year, just a few blocks from the city centre, Kinglet offers fine dining from its late-1800s premises that was once a tree-processing mill. Named after the tiny but tenacious kinglet bird found in the Blue Mountains that flank the Walla Walla Valley, the restaurant's focus is on seasonal, local and sustainable ingredients. Reserve seats at the communal chef's counter and enjoy the five-course tasting menu, with or without wine pairings. Open Thursday-Monday, from 4.30pm. **kingletww.com**

Seasonal dining at Kinglet

PASSATEMPO
215 W Main Street, Walla Walla, WA 99362

With its retro-style booths and 1950s-era cocktail glasses, Passatempo exudes a casual neighbourhood Italian ristorante feel. Dig into a plate of handmade pasta, gnocchi or slice of sourdough pizza, made using ingredients sourced from local farms and speciality purveyors. Highlights include the pork and prosciutto meatballs, carbonara pasta made with organic farm-fresh eggs, and rigatoni alla Bolognese. Open Thursday-Monday, 4pm-9.30pm. **passatempowallawalla.com**

WALLA WALLA VALLEY: BEYOND WINE

While wine is the primary drawcard of the Walla Walla Valley, there's much more for visitors to do. For active, outdoors-y types there's skiing and snowshoeing during winter, and in warmer months there's cycling (or renting electric bikes), kayaking or bird watching. Downtown Walla Walla has also become a haven for artists, with a number of studios, galleries and sculptures dotted around the city and the **Whitman College campus**. Join a guided tour with **First Friday Art Tours** *(artwalla.com)*, held each month. The region is also home to social and educational wine event **Celebrate Walla Walla Valley Wine** *(wallawallawine.com/celebrate)* in July, and, in late October, the **Walla Walla Balloon Stampede (see Facebook).**

VANCOUVER ISLAND

Lying just off British Columbia's west coast, this Pacific-facing island may not be Canada's best-known wine tourism draw, but it offers a bounty of hidden gems. And vying for the accolade of Canada's warmest place, it's the perfect destination for a winter getaway.

View of the Inner Harbour at sunset in downtown Victoria

Only a 90-minute ferry ride from Vancouver in Canada's western province of British Columbia (BC), or a 50-minute seaplane journey from Seattle in the US, a lush isle beckons with artisanal wineries and a unique climate. It's Vancouver Island, a relatively unknown wine appellation, but a charming and convenient destination for enthusiasts.

While Vancouver Island stretches 460km north to south, most winemaking happens in a small area along its southeast coast – the Cowichan Valley, the island's only GI sub-appellation, which surrounds the city of Duncan and extends between Mill Bay northwest up to Cowichan lake (see maps, p158). Most wineries are on the east coast, off the Saanich Inlet.

A short drive from Victoria – the island's hub city and BC's provincial capital – the Cowichan Valley is a breeze to navigate. The short distance between wineries – at most 20 minutes' drive apart – allows for leisurely exploration.

'Winemaking' on the island dates back 100 years, first using loganberries (a blackberry-raspberry cross) to create fruit wines. The focus turned to grapes in the 1980s with the Duncan Project, a government-led test site initiative. Over seven years, more than 100 grape varieties were trialled before the government withdrew funding. Today, a vibrant community of winemakers produce a diverse array of grapes that reflect the island's individual microclimates.

Warm welcome

Nearby mountains shield the Cowichan Valley from Pacific ocean storms from the west. The First Nations Salish translation of Cowichan is 'warm land', epitomising the area's long, dry growing season with low frost risk. Thanks to a year-round average temperature warmer than anywhere else in Canada, the island's climate is known as 'Maritime Mediterranean'.

Hybrid varieties such as Ortega, Auxerrois and Maréchal Foch were the focus in the early days, while today – thanks in part to

climate change – some Vitis vinifera flourish, particularly Pinot Noir and Pinot Gris. However, Chardonnay will soon be the most planted white grape variety.

This potential has attracted international investment. California-based Jackson Family Wines (JFW) owns two Cowichan Valley properties, accounting for 53ha of the island's area under vine: 130ha according to the BC Wine Grape Council's report for 2022.

JFW family members bought Unsworth *(unsworthvineyards.com)* in 2020, focusing on sustainable vineyard practices. Disease-resistant cross-varieties developed by Swiss geneticist Valentin Blattner since the 1980s, such as Petite Milo, Cabernet Libre and Labelle, play an important role in Unsworth's signature white and red blends Allegro and Symphony respectively. Its Charme de L'Ile sparkling wines (a trademarked concept used by Vancouver Island producers who make Charmat-method wines) are BC's most accessible expressions of the style.

Unsworth's Charme de L'Ile sparkling Pinot Noir rosé

In 2022, JFW also acquired Blue Grouse (bluegrouse.ca), one of Cowichan Valley's pioneering wineries. Its portfolio centres around Chardonnay, Pinot Noir and 30-year-old dry-farmed Pinot Gris. Annual production is about 84,000 bottles, set to increase to 120,000.

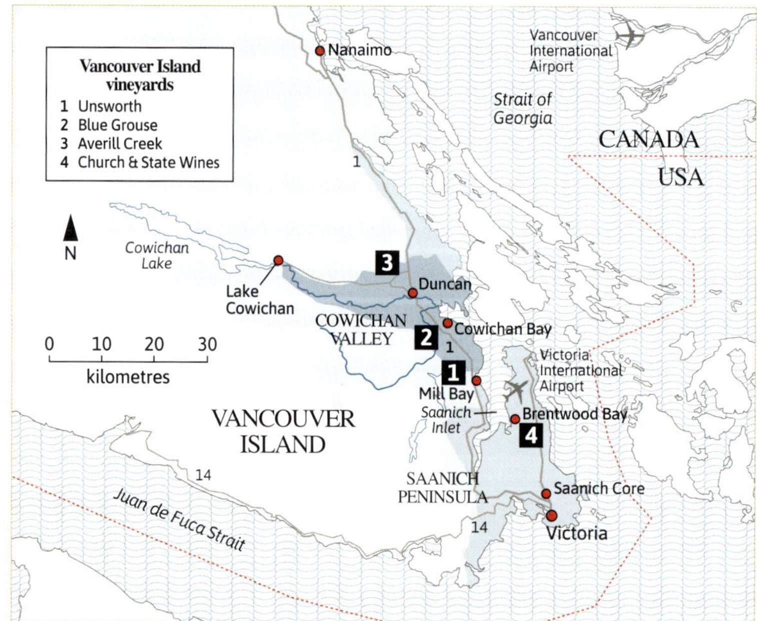

Plan ahead

While most of Vancouver Island's wineries offer a year-round visiting experience, it's worth planning ahead in the off-season, as some limit their tasting hours. At the same time, it's common for restaurants to close on Mondays and Tuesdays. Averill Creek *(averillcreek.ca)*, about 12km from the coast, is a must-visit in winter when it offers its popular Alpine cheese fondue and wine pairing experience (CA$50/£29 + tax per person, reservations required). Its tasting room and outdoor terrace (when weather permits) offer

GETTING THERE

There are frequent flights to Victoria International (at Sidney) from Canada's major cities (such as Vancouver: 30 mins, from $106/£64), but renting a car is necessary to visit wineries. Several daily ferries run from Vancouver (1hr 35mins into Swartz Bay, adults CA$17.20/£10). From Seattle, consider the seaplane route (50-75 mins, from about $200/£117 per person) or the Clipper ferry (2hrs 45mins, from about $99/£58 per person).

Viewable from Malahat Skywalk, the impressive volcanic peak of Mount Baker lies almost as far northwest as you can get in the USA

MY PERFECT DAY IN VANCOUVER ISLAND

MORNING & LUNCH

Wake up to the gentle sounds of waves at Oceanfront Suites* in the charming community of Cowichan Bay. Stroll to nearby **Leeward Cafe** *(@leewardcafe)* for a quick breakfast sandwich and honey-sweetened chai latte before browsing in the waterfront shops or, for the adventurous, taking to the water with **Cowichan Bay Kayaking** *(+1 250 597 3031)*. In your rental car, head to Averill Creek (see p158) to enjoy one of three tastings that you have already booked – guided, self-guided or the full package, including private vineyard and cellar tour. You could stay and order from the snack menu, or head where the locals go: **Hank's Cowichan (hankscowichan.com),** about 15 minutes' drive away and a great sit-down stop for a homemade soup and sandwich. Alternatively from Hank's, grab some **Pickles' Pantry** *(picklespantry.ca)* paté or charcuterie items for an on-the-go snack — also to be found at the Duncan Farmers Market on Saturdays.

AFTERNOON

After a bite, continue to **Blue Grouse Estate Winery** (p158), only eight minutes' drive from Cowichan. The modern tasting room is inviting and boasts panoramic views of the vineyards. From here, a 30-minute drive south gets you to the **Malahat Skywalk***, where a spiral wooden ramp leads up to a 32m viewing deck. Soak up the breathtaking scenery of Mount Baker (above) and the Saanich Peninsula before taking the nine-second spiral slide to the bottom. Then make your way to the Mill Bay ferry to cross the inlet. The 25-minute journey runs nearly every hour until 6:30pm (from CA$32.60/£19 for car with two passengers).

EVENING

After docking at Brentwood Bay, drive 10 minutes to **Church & State** (p160) to taste traditional-method artisan sparkling wines. Afterwards, check in and relax at **Brentwood Bay Resort & Spa***. Dinner options include the pub (a great chowder of local fish and clams) or the upscale **Arbutus Room**, where you can savour seared Hokkaido scallops or Saltspring Island mussels.

For more details of entries marked with an asterisk (*), see p161

Cowichan Bay Kayaking

Whale watching in Cowichan Bay (see 'Address book', p161)

stunning panorama views of its 13ha of vineyards on the sunny southeast-facing slopes of Mount Prevost. It's the perfect spot to sip Averill Creek's Joue Red, a blend of Foch, Gamay and Pinot Noir, or its wild-ferment Westholm Pinot Gris.

At family-run Rocky Creek *(rockycreekwinery.ca)*, in tourist season 30- to 45-minute tastings are available at a modest charge, and spring and summer are best for enjoying an informal picnic on the patio with wine purchased. But even during winter you can still enjoy a stroll through the gardens next to its estate vineyards before enjoying a local Cabernet Foch or wines from international varieties.

While most of Vancouver Island's wine comes from Cowichan Valley, there are vineyards elsewhere, such as in the Saanich Peninsula, where Church & State Wines *(churchandstatewines.com)* has its satellite property. Best known for producing elegant whites and bold reds from the southern Okanagan Valley (mainland BC), Church & State's 8ha island estate (about 5ha planted) is focused on sparkling wines. Try the traditional-method, hand-riddled Gris de Noir, which comes from Saanich-grown Pinot Noir.

Taste of the island

Vancouver Island boasts a robust culinary identity that draws foodies as well as wine lovers. Organic produce abounds and menus feature freshly caught Pacific salmon and succulent Dungeness crab, while farm-to-table dining showcases the island's bounty of artisanal cheeses and grass-fed meats.

Wine festivals add another layer of allure. Averill Creek's Noir Fest, in late June, celebrates Pinot Noir from across BC *(averillcreek.ca/noir-fest)*. In the warmth and full bloom of August, the Cowichan Valley Wine Festival *(@cowichanwineries)* spotlights local wines, dishes and live entertainment. Victoria Wine Festival *(vicwf.com)* wraps up the season around harvest time in late September, with winery dinners and events aimed at both consumers and wine trade.

Whatever the season, and whether you're a seasoned oenophile or a casual sipper, Vancouver Island promises unforgettable experiences.

Church & State Wines (see left)

The spiral staircase and accessible viewing ramps of the Malahat Skywalk tower, just south of Mill Bay

YOUR VANCOUVER ISLAND ADDRESS BOOK

ACCOMMODATION

Brentwood Bay Resort This waterfront sanctuary offers luxurious accommodations, spa indulgence and panoramic views. It's a tranquil escape where elegance meets natural beauty. **brentwoodbayresort.com**

Oceanfront Suites at Cowichan Bay Perfect for nature lovers, Cowichan Bay is an idyllic and historic waterfront community with breathtaking views, known for its independent shops and cafés. **oceanfrontcowichanbay.com**

Parkside Hotel & Spa A luxurious central spot in Victoria to start or end your trip, blending modern elegance with environmental sustainability in its spacious suites and rooftop gardens. **parksidevictoria.com**

FOOD & DRINK

Alpina Restaurant at Villa Eyrie Resort For menus that infuse local ingredients with traditional Alpine European flavours, accompanied by breathtaking 270° views of the Olympic mountains, Mount Baker and the Saanich Inlet, this is a dining experience elevated in every aspect. **villaeyrie.com/alpina-restaurant**

Little Jumbo A must-visit in Victoria for gastronomic adventurers, with an intimate setting and inventive menu that showcases locally sourced ingredients and craft cocktails. **littlejumbo.ca**

The Lakehouse at Shawnigan A serene hideaway with views overlooking the lake. It supports local distilleries, cideries, breweries, wineries and farms, so the menu always features seasonal dishes. **atthelakehouse.ca**

Alpina Restaurant at Villa Eyrie Resort

THINGS TO DO

Malahat Skywalk An awe-inspiring treetop adventure en route to Cowichan Bay, offering panoramic views of forests and mountains for an unforgettable nature experience (CA$36.95/£22 per adult). **malahatskywalk.com**

Victoria Harbour Victoria's downtown area features iconic attractions such as the Parliament Buildings and historic Empress Hotel, plus vibrant street performers along Inner Harbour. Enjoy scenic walks, boat tours and charming waterfront cafes. **tourismvictoria.com**

Whale watching At eco-friendly business Ocean Ecoventures in Cowichan Bay, get up close and personal with marine life (pictured, p160) with expert guides: a thrilling exploration of the ocean's wonders (Cowichan Bay Whale Watching Eco Tour, half day CA$169/£99 per adult). **oceanecoventures.com**

REST OF THE WORLD

Choose your wine style, and your backdrop – what's it to be? The inky reds and soaring peaks of Mendoza, coastal vineyards in Marlborough, famous for its zingy Sauvignon Blancs, or amphora-aged wines from historic Georgia? And how long have you got? In five days you can tour three wine regions from Argentina's historic city of Córdoba, or take a 5km drive in Australia's Margaret River and enjoy its elegant Cabernets and Chardonnays. The world is your oyster ...

Vasse Felix vineyard and winery

MARGARET RIVER

This 5km driving route is an ideal starting point to explore Margaret River, Western Australia's most famous wine region. Resident Danielle Costley reveals a kaleidoscope of superb tastes and experiences along the way.

Spectacular coastlines, lush forest, valleys blanketed with grapevines – it all awaits in Western Australia's Margaret River region, where winemaking is its lifeblood and ancient soils nourish its soul. Some of the world's finest Chardonnays and Cabernets are produced here, in a region that celebrates its wines as much as the abundance of fresh produce, especially in November during the annual Gourmet Escape wine and food festival (usually in November, visit *(gourmetescape.com.au)*.

While its origins date back to the 1920s, with the planting of an Italian grape variety called Fragola, it wasn't until 1967 – when Dr Tom Cullity established Vasse Felix winery – that the Margaret River wine region was truly born. Cullity's humble enterprise began with just 3ha of land and plantings of Cabernet Sauvignon, Shiraz, Malbec and Riesling.

Times have certainly changed since then, with 175 wineries now established throughout the region, many with restaurants that feature local produce as the dish of the day.

The Tom Cullity Wine Trail *(margaretriver.com)* pays homage to this vigneron and cardiologist. Stay in Margaret River (300km south of Perth) and take a day or two to drive along the 5km stretch of Tom Cullity Drive that is dotted with family-run wineries, wildflowers during spring and a gushing Wilyabrup Creek in the winter months. Stop for a glass of wine, a waterside picnic or a fine-dining lunch, and discover Margaret River's evolution from humble beginnings.

Begin your journey at The Margaret River Chocolate Company *(chocolatefactory.com.au)*, where you will find more than 200 chocolate products, including truffles made

FACT FILE: MARGARET RIVER

AREA UNDER VINE 5,480ha, accounting for 2% of Australia's wine production and 25% of Australia's premium wine production
VARIETIES PLANTED 36 **WINES PRODUCERS** 175 **CELLAR DOORS** 90

'This region celebrates its wines as much as the abundance of fresh produce'

Heydon Estate

with gin and tonic or salted caramel, as well as macadamia nut clusters, gourmet chocolate bars and chocolate-coated honeycomb.

Fill your picnic basket at Providore *(providore.com.au)*, a gourmet deli stacked with homemade tapenades, jams, chutneys, cured meats, dressings, cheeses, olive oil, wine and liqueurs. Everything is made on site – even the cold-pressed extra virgin olive oil from its Tuscan olive trees, with most produce picked from its organic vegetable garden.

A visit to the cellar door at Heydon Estate *(heydonestate.com.au)* reveals the family's passion for cricket, with wine labels such as The Urn and Hallowed Turf. Try the Cherry Viognier Rosé or The Willow Chardonnay.

You'll be surrounded by French oak barriques in a contemporary barrel room when stopping by for a tasting at Thompson Estate *(thompsonestate.com)*. Established by Peter Thompson in 1994, the Four Chambers label acknowledges his – and Tom Cullity's – parallel careers of vigneron and cardiologist.

The picturesque Juniper Estate *(juniperestate.com.au)* is situated on the banks of Wilyabrup Creek. The rustic cellar door offers a good selection of varietal wines from winemaker Mark Messenger. A quiet achiever, the winery is renowned for its award-winning reds, but the Fianos are equally impressive. Keep an eye out for the rare Baudin's black cockatoos that live on the property.

Follow the winding blue gum- and jarrah-lined road until you arrive at a mud-brick cellar door. This colonial-style tasting room was built by the Devitt family from the mud in its dam when they founded Ashbrook Estate *(ashbrookwines.com.au)* 40 years ago. Riesling was the most-planted variety in Margaret River then, and while Chardonnay and Cabernet are

GETTING THERE

Fly direct to Perth, Western Australia's capital. From there, hire a car and drive the 300km, three-hour drive south to Margaret River. Or book a Cessna turbo-prop seaplane or helicopter and be there in an hour.

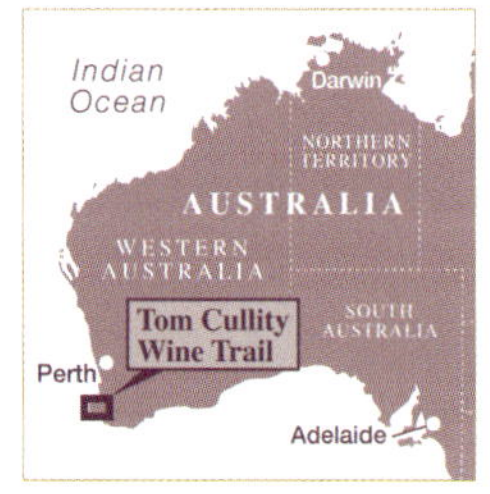

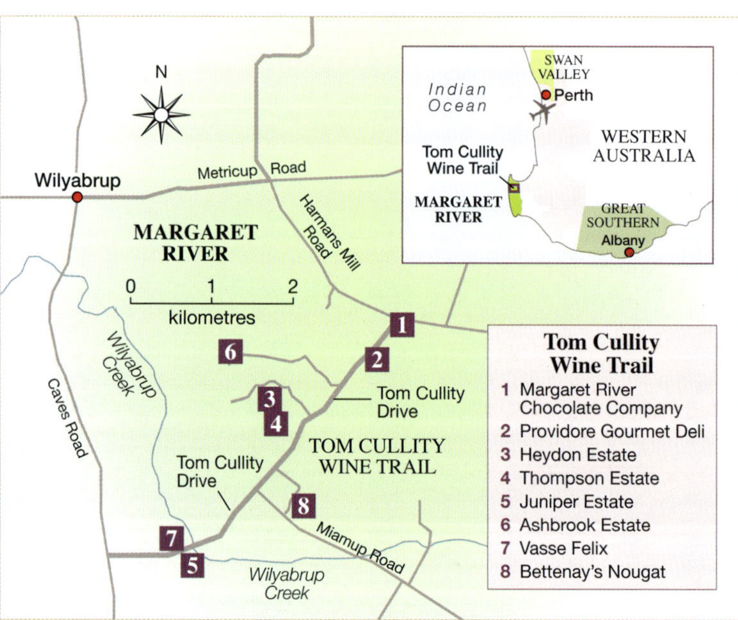

Domaine Naturaliste

MY PERFECT DAY IN MARGARET RIVER

MORNING

Stay at **Cape Lodge** *(capelodge.com.au)*, a stylish country house hotel. Opt for an inclusive experience, which incorporates breakfast, afternoon tea and a vineyard tour. Rooms are light and spacious, offering garden or lake views.

After breakfast at the lodge, drive south for 20 minutes to taste your way through some of the region's finest Chardonnays, Cabernets and more at **Xanadu Wines** *(xanaduwines.com)*.

LUNCH & AFTERNOON

Not far from here is **Voyager Estate** *(voyagerestate.com.au)*, where your sensory exploration of the region continues with a four- or seven-course gourmand feast created by Galician-born head chef Santiago Fernandez, and inspired by the estate's wines. Plating is an artwork, each dish featuring the native botanicals of the region.

On the return trip to your accommodation, stop at **Domaine Naturaliste** *(domainenaturaliste.com.au)*, whose cellar door opens directly onto its vineyard. Try a wine flight paired with cheeses or a charcuterie plate. Arrive back at Cape Lodge for afternoon tea, followed by an escorted vineyard tour and wine experience.

EVENING

Enjoy sunset drinks on the lakeside deck and a three- or five-course Trust the Chef menu. The restaurant's modern French bistro fare is infused with fresh Western Australian seafood, accompanied by a wine list featuring local and French varieties.

Tony Howell, chef at Cape Lodge

Xanadu Wines

CLOCKWISE FROM TOP Voyager Estate; rare Baudin's black cockatoos, as seen at Juniper Estate; Settlers Tavern; Vasse Felix's Tom Cullity Cabernet Sauvignon-Malbec; Ashbrook Estate cellar door; the cheese platter at Providore

'Stop for a glass of wine, a waterside picnic or a fine-dining lunch, and discover Margaret River's evolution from humble beginnings'

the hallmark varieties of the region today, the Devitts continue to produce a Riesling from those original vines.

At Vasse Felix *(vassefelix.com.au)*, toast the man who saw the potential of this region with a glass of the Tom Cullity Cabernet Sauvignon-Malbec. Margaret River's wine journey began from those grapes, picked from the founding vineyard.

While here, indulge in Brendan Pratt's Japanese- and Korean-infused fine-dining cuisine, including kingfish wing in a smoked eel and mushroom XO sauce partnered with the Heytesbury Chardonnay. Or book a Cellar Experience tour with a stroll through the organic vines, before a back-vintage tasting in The Vault, which houses Tom Cullity's first bottle of Cabernet Sauvignon-Malbec from its first vintage in 1972.

You've come to the end of the trail, but finish on a sweet note at Bettenay's Nougat *(bettenaysmargaretriver.com.au)*. Watch the confectionery being made while enjoying a piece or two and sipping on a honey liqueur.

Olio Bello

YOUR MARGARET RIVER ADDRESS BOOK

ACCOMMODATION

Constellation Apartments Stay on the town's main street at these stylish one-bedroom flats, complete with modern conveniences and original artworks. **constellationapartments.com.au**

Olio Bello Glamp in a safari-style lakeside bungalow at this organic olive grove. Stroll to the café next door for an olive oil tasting, lunch and glass of wine. Breakfast hampers and dinner platters are also available. **oliobello.com**

Pullman Resort Let the ocean lull you to sleep at this resort on Bunker Bay where a boardwalk leads to the beach. Choices range from studio rooms to family bungalows. **pullmanbunkerbayresort.com.au**

Bunkers Beach House

RESTAURANTS

Bunkers Beach House Dine on sustainably caught local seafood in this absolute beachfront dining. Dishes change daily but can include roast whole nannygai fish or dry-aged amberjack carpaccio. **bunkersbeachhouse.com.au**

La Scarpetta Trattoria Enjoy traditional Italian cuisine and homemade pasta in a relaxed vintage-chic setting, with an extensive Italian and local wine list. **lascarpetta.com.au**

Pizzica Authentic wood-fired Italian pizzas and charcoal-grilled t-bone steaks, pork ribs and lamb chops in this humble, yet welcoming rustic pizzeria. **pizzica.com.au**

Yarri Drive 45km north to Dunsborough and tuck into authentic Australian cuisine, where locally foraged native foods star on this ever-changing menu. **yarri.com.au**

BARS

River Hotel A favourite watering hole for locals to while away an afternoon in the beer garden with a selection of locally crafted spirits and beer, accompanied by live music or a silent disco. **theriverhotel.com.au**

Settlers Tavern With 600 wines on its award-winning list, the tavern also offers small pours of select back-vintage wines via the Coravin preservation system. **settlerstavern.com**

SHOPS & TOURS

Margaret River Collaborative Here, several artisanal producers specialising in local, handmade products are showcased under the one roof. There's coffee, tea, crafts, clothing, cheese, artworks, body products and wine. **margaret-river-collaborative.business.site**

Walk Talk Taste Bring your appetite on this walking brunch tour and discover some of the region's best local produce. Sip on cold-drip coffee with chocolate, freshly made ice cream from a local dairy, pair cheeses with wine, and taste freshwater marron (crayfish). **walktalktaste.com**

MARLBOROUGH

Home to New Zealand's pioneering 1973 Sauvignon Blanc plantings, this region at the top of South Island is well worth a stop for its breathtaking scenery alone. Take a journey through Marlborough's valleys, full of well-known and hospitable wineries.

Picton town and its sheltered harbour for the Cook Strait ferry

Marlborough's problem is one of transition: for people visiting New Zealand, even for many nationals, it is a place you go past if you're going somewhere else. Its location – a 30-minute drive south of Picton (the main port of entry to the South Island via the Cook Strait car ferry from Wellington) – means people are either trying to make it north to the ferry or continuing south to Christchurch, Otago or even west to Nelson and the West Coast region.

That's a shame because Marlborough has everything to offer if you just stop and take a bit more time. Its main centre Blenheim is a functional town but set in beautiful part of New Zealand. And the mountains and hills feel close, especially the dramatic Richmond Range to the north.

'Time to head into the locals' playground: the stunning Marlborough Sounds'

Indeed, as Auntsfield winemaker Luc Cowley pointed out, you can always orient yourself in the Wairau Valley, Marlborough's main wine-growing area: the blue mountains (the Richmond Ranges) lie to the north and the green mountains lie to the south. These latter are more a shade of brown in summer, hence their name: the Wither Hills.

The Marlborough Sounds, too – a collection of ancient valleys flooded with Pacific ocean waters along a 1,500km stretch of coastline – is a stone's throw away. While many patrons of the car ferry admire its beauty as they pass through the parade of pristine, almost uninhabited bays and coves before hopping in their vehicle and driving on, it is a very good reason to hang around.

Diversity in abundance

There is much else on offer in Marlborough – and that is true of the wines, too. The region has considerably more to recommend it than the Sauvignon Blanc upon which it made its reputation. What's more, most cellar doors are focused in a relatively small

Cloudy Bay's Barracks vineyard in Omaka Valley (see p173)

'Most cellar doors are focused in a relatively small and easily navigable area around Blenheim and Renwick to its west'

Marlborough winery self-guided biking tour

and easily navigable area around Blenheim and Renwick 8km to its west, with the region's airport lying in between.

In this road trip, we've allocated two days to the Wairau Valley so visitors can really get an idea of how diverse the sub-region and its wines can be. Further information can be found on the wine tourism map at *marlboroughwinenz.com*, while the more energetic traveller can take advantage of local bicycle routes.

We've grouped our three-day itinerary so it can easily be rearranged. For instance, those travelling down from North Island could arrive in Picton and head straight to the Marlborough Sounds before returning to Picton the next morning and heading south to Blenheim. Both days 1 and 2 finish around Renwick, which allows travellers to take the 1.5-hour drive along State Highway 6 (SH6) further west to Nelson.

If you only have two days available, start with the day 2 itinerary then do day 1's suggested trip in reverse (ie, drive out to Renwick and head up SH6 before turning east onto Rapaura Road). Fit in the wineries you want before continuing on Rapaura Road to meet SH1 and continue your journey south.

DAY 1
Wairau Valley from Blenheim

Grab a breakfast coffee and toastie from Sammies on Scott Street. A Kimcheese (kimchi and cheese – add a pork and fennel sausage pattie if you want) is a monumental start to anyone's day. From there, it's a short drive just out of town to Lawson's Dry Hills *(lawsonsdryhills.co.nz)* – a pioneer of sustainable practices in Marlborough and producer of fine aromatic white wines. From here, a variation on our itinerary would be a 15-minute drive down State Highway One (SH1) to Seddon and the Awatere Valley. This sub-region is slightly cooler than the main Wairau Valley and contains many newer – and sizeable – vineyard plantings, which deliver some truly exciting wines. One of the few cellar doors here is Yealands *(yealands.co.nz)*.

But our main route heads north on SH1 for 15 minutes, crossing the Wairau river to stop off the roundabout at Tuamarina to visit the memorial to the Wairau Affray of 1843 and a glimpse of the country's often violent colonial past.

From Tuamarina, head back the way you came and turn right onto Rapaura Road at Spring Creek where a number of wineries have their cellar doors. We suggest the pan-regional Saint Clair *(saintclair.co.nz)* and idiosyncratic Rock Ferry *(rockferry.co.nz)* on the way to a tasting and lunch at Cloudy Bay *(cloudybay.com)* on Jacksons Road. One of Marlborough's flagship brands, owned by French luxury goods giant LVMH, its cellar door will impress even the most corporate-cynical.

RIGHT FROM TOP Nautilus Estate's main entrance and cellar door on Rapaura Road, Renwick. Te Whare Ra owner-winemakers Anna and Jason Flowerday in among the buckwheat

From Cloudy Bay (pictured, p172), get back onto Rapaura Road and head west (and then south) to Renwick. There are many wineries on this stretch, so how many you have time for will depend on how long that lunch was.

The proudly organic Te Whare Ra *(twrwines.co.nz)* is one of Marlborough's must-visit wineries and Framingham *(framingham.co.nz)* has a well-deserved reputation for its Rieslings. Hans Herzog *(herzog.co.nz)* produces a range of unexpected varieties and wines, while Huia *(huiavineyards.com)* is an unsung gem. Bubbles fans can stop in at No1 Family Estate *(no1familyestate.co.nz)*, while Nautilus *(nautilusestate.com)* provides a glimpse into day 2 as some of its fruit (in particular its Pinot Noir) is sourced from the Southern Valleys sub-region. And the excellent Fromm *(frommwinery.co.nz)* is near the airport on your 11-minute drive back to Blenheim.

DAY 2

Southern Valleys from Blenheim

Fuel up for the day at Burleigh Gourmet Pies, a Marlborough landmark on the southwestern edge of Blenheim town – try the signature pork belly pie. Then, just a few hundred metres west along New Renwick Road is the turn-off for the Omaka Aviation Heritage Centre (see 'Address book', p175), which includes Sir Peter Jackson's impressive collection of World War I aircraft.

Back on New Renwick Road and the turn-off south to Auntsfield *(auntsfield.co.nz)* marks the first stop on our Southern Valleys wine tour where wind-blown loess has settled in the valleys and provided the clay so essential to its Pinot Noir.

Brancott Valley is the next one along and, although there are no major cellar doors, a number of producers source fruit from here, and Dog Point *(dogpoint.co.nz)* sits at the base of the ridge that separates Brancott Valley from Omaka Valley. At the head of the Brancott Valley lie two vineyards with

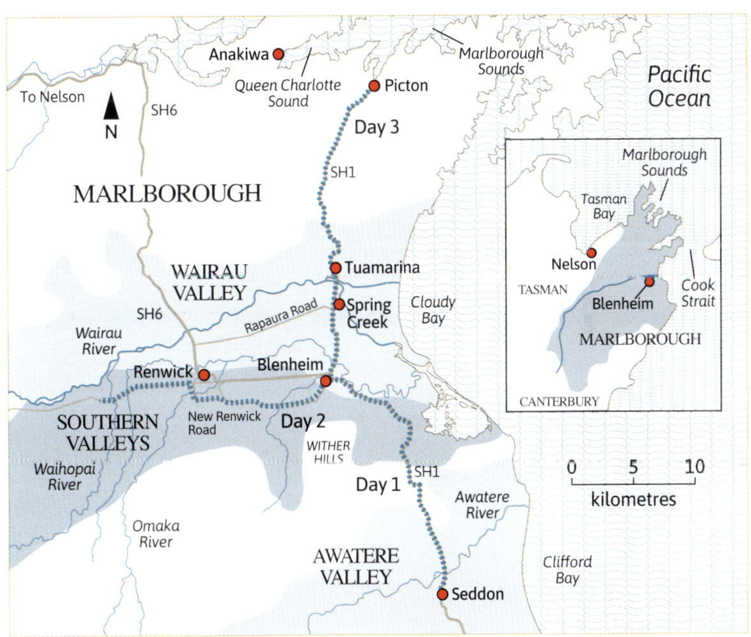

GETTING THERE

If you already have your own vehicle, Cook Strait ferries run between Wellington and Picton, the journey time being around 3.5 hours; otherwise it's a 300km (4-4.5 hour) scenic drive from Christchurch. Otherwise, hire a car after arriving into Marlborough Airport Blenheim: Air New Zealand flies from Auckland and Wellington, and Sounds Air runs limited services from Christchurch, Wellington and Paraparaumu.

Winemaker Jules Taylor under bird netting at The Wrekin vineyard in Brancott Valley

Bay of Many Coves Resort

stellar reputations: Clayvin, now part of Giesen *(giesen.co.nz)*, and The Wrekin *(wrekin.co.nz)*. The latter's excellent organic fruit is used by numerous brands and highlighted on their wine labels, from one-man-band outfits to the more established Jules Taylor Wines *(julestaylor.com)*.

Back onto the New Renwick Road again and continuing onto Dog Point Road, Omaka Valley is the next one along, and here you'll find the must-visit Greywacke *(greywacke.com)* – make sure you book an appointment-only, no-charge tasting in advance (weekdays 11am or 2pm) by emailing through the website.

Drive back up towards Renwick to take the SH63 west a short distance before ducking down to the more expansive Waihopai Valley. Once home to the New Zealand government's not-so-secret satellite listening facility (the huge white domes have since been dismantled, although the facility is still in operation), it also houses Spy Valley Wines *(spyvalleywine.co.nz)* and Churton – both well worth a visit, although Churton is by appointment only (contact the winery directly via the website: *churtonwines.co.nz*).

Finally, call in at nearby Clos Henri *(closhenri.com)*, which is owned by the Bourgeois family from Sancerre in the Loire, before heading back to Blenheim.

DAY 3

Picton, Nelson, Marlborough Sounds

Time to head into the locals' playground: the stunning Marlborough Sounds. Drive up to Picton with a quick stop at Johanneshof cellar door *(johanneshof.co.nz)* on the way. Once in town, park the car, check the ferry timetable and stop in at Toastie *(toastie.co.nz)* for a snack while you wait.

There is a range of truly breathtaking resorts and lodges throughout the Sounds, each of which offers a variety of activities, but Punga Cove, Bay of Many Coves Resort and Furneaux Lodge are among those recommended.

The more adventurous early-riser might want to combine this leg with a spot of guided kayaking or a stint on the Queen Charlotte Track, a 71km-long walking and biking track running from Anakiwa, at the head of Queen Charlotte Sound in the south (about 30 minutes' drive from Picton) to Meretoto/Ship Cove in the north of the Sounds. A local operator such as Wilderness Tours *(wildernessguidesnz.com)* should steer you in the right direction.

Once you've chosen your activity, board the ferry, or bike, walk or paddle your way to your chosen destination. Then, whether you've hiked, paddled, snorkeled, sailed or cycled – or merely strolled the jetty – it's time to sit back and relax with a glass of Marlborough wine.

Marlborough Wine & Food Festival

YOUR MARLBOROUGH ADDRESS BOOK

ACCOMMODATION

14th Lane Ideally located in central Blenheim. Formerly The Builder's Arms, the views aren't great and it boasts no restaurant or bar, but the rooms more than compensate. **14thlane.nz**

Furneaux Lodge Tucked all the way up in the stunning Marlborough Sounds, just over the hill from Captain Cook's favourite NZ anchorage, this place is as magical in the rain as it is in the sun. **furneauxlodge.co.nz**

Hotel d'Urville Full of character, both inside and out, the rear bar flows into an outside area which doesn't quite match the inside vibe but is surprisingly comfortable on summer evenings. **hoteldurville.com**

RESTAURANTS & CAFES

Arbour The fine-dining Marlborough experience on everyone's lips, winemakers included. Named NZ's Best Regional Restaurant in 2022 by Cuisine, it also has a coveted 'two hats' rating from the food and drink lifestyle publication. **arbour.co.nz**

Frank's Oyster Bar and Eatery Another restaurant that comes recommended by local winemakers, you'll find this one in central Blenheim. **eatatfranks.co.nz**

The Store In Kekerengu, a 50-minute drive south of Blenheim on SH1, its sweeping vistas across a wild coastline make it well worth the stop on the way to Christchurch. **thestore.kiwi**

Furneaux Lodge

THINGS TO DO

Festivals The summer months of January, February and March respectively feature the Picton Maritime Festival, the Marlborough Wine & Food Festival and the Havelock Mussel and Seafood Festival. **maritimefestivalpicton.com marlboroughwinefestival.com havelockmusselfestival.co.nz**

Marlborough Farmers Market Held on Sundays at Blenheim's A&P showgrounds, this is a great way to enjoy a range of regional produce with a minimum of travel. **marlboroughfarmersmarket.org.nz**

Omaka Aviation Heritage Centre A must-visit for historic aircraft enthusiasts, or for an extra-special occasion, this renowned aviation museum at Blenheim offers the chance to book a flight on a fully restored World War II-era Avro Anson – a package starting at NZ$990 (£475) per person including a 25-minute flight. Or a 20-minute flight in a Boeing Stearman biplane, costing NZ$395 (£189) for one or two people. **omaka.org.nz**

CÓRDOBA

We head to Argentina's second largest city and its surroundings in the centre of the country, with a dream itinerary for a five-day tour introducing three very diverse wine regions.

Plaza San Martín and the cathedral in the centre of Córdoba city

Córdoba often plays third fiddle to Argentina's glamorous capital Buenos Aires and the country's key wine city Mendoza, but this picturesque province is a surprisingly diverse destination.

A rich 16th-century history, tracts of rolling countryside and hospitable Fernandito*-sipping cordobeses have long ensured its reputation as a favourite with domestic travellers (*the local favourite mixed drink of Fernet-Branca amaro with cola). And now Córdoba is making its mark as a wine destination. While the 1573-founded eponymous provincial capital, with its UNESCO-heritage Jesuit Block and buoyant nightlife, is a fascinating cultural introduction, Córdoba's winemaking regions make for an exciting road trip. Sierras and mountains paint a dramatic landscape, replete with rivers, cattle ranches, and opportunities for outdoor adventures such as parasailing, horse riding and hiking... Like Mendoza (750m and up), Córdoba province is elevated (about 350m-550m) but, at about 470km to the northeast of Mendoza, is under a less intense Andean gaze.

Jesuit and Spanish colonial history weave a colourful architectural tapestry, the perfect backdrop to the region's 400-year-old story of winemaking that began with sacramental wine. In its heyday, the Sierras Chicas hills were home to 1,500ha of vineyards, and while only 277ha are cultivated by 20

FACT FILE: CÓRDOBA

VINEYARDS PLANTED 277ha
WINERIES 20 bodegas across five regions
REGIONS (FIVE) Sierras Chicas, Valle de Calamuchita, Traslasierra, Punilla, Norte Córdobes
MAIN GRAPES *White* Sauvignon Blanc, Chardonnay, Viognier *Red* Isabella, Pinot Noir, Malbec, Merlot, Ancellotta, Tannat
TOURS Rutur *(ruturviajes.com.ar)*
CAR HIRE Sixt (+54 351 569 4310)
MORE INFO cordobaturismo.gov.ar (click 'Qué hacer', then 'Caminos del Vino')

bodegas in three key regions today, it represents a shift to quality over abundance. A prevalent Germanic culture means Córdoba province also hosts Latin America's largest Oktoberfest, backed by a dynamic craft beer scene to perk up tannin-saturated palates.

DAY 1

Sierras Chicas & Colonia Caroya

Córdoba's wine story begins a 40-minute drive north of the provincial capital in Colonia Caroya. It's an ideal day trip, but better extended with an overnight estancia stay. Take the scenic, slightly longer route through quaint villages that open up to peach and fig orchards, and onto vineyards.

The Jesuits constructed estancias here and in nearby Jesús María in 1616 and 1618 respectively, beacons on the viceroyalty's Camino Real (Royal Route) to Buenos Aires and the Río de la Plata, from where the first sacramental wine, produced in Colonia Caroya, set sail to cross the Atlantic for Felipe V's sipping pleasure. When Italian immigrants from Veneto and Friuli settled here from around 1878, given land in return for work, their agricultural (and charcuterie) know-how and their introduction of Vitis labrusca Isabella helped write the next chapter. Ancellotta landed a century later, cultivated first in Caroya before the Zuccardi family took the variety to Mendoza.

Visits to Terra Camiare *(terracamiare.com)* and La Caroyense *(bodegalacaroyense.com.ar)*, founded in 1928 and 1930, recount Caroya's past and future. Oenologist Gaby Campana contributes to setting the province's winemaking standard at Terra Camiare, while the magnitude of the 505hl vats at La Caroyense showcases the area's bygone winemaking muscle. Fourth-generation vintner Gaby plays around with old Isabella and Pinot Noir (known as la francesa, 'the French one') vines. He's a pioneer who microvinifies Ancellotta, Sauvignon Blanc and Chardonnay in concrete eggs. His top-line Socavones Capitulum Semillon comes from Quilino, a desert-like area where summer temperatures can reach 45°C.

Other Quilino projects include Piensa Wines *(bodegapiensa.com.ar)*, led by Bordeaux-based Alejandro López, whose Cabernet Sauvignon-Cabernet Franc Reserva adds local DNA with small

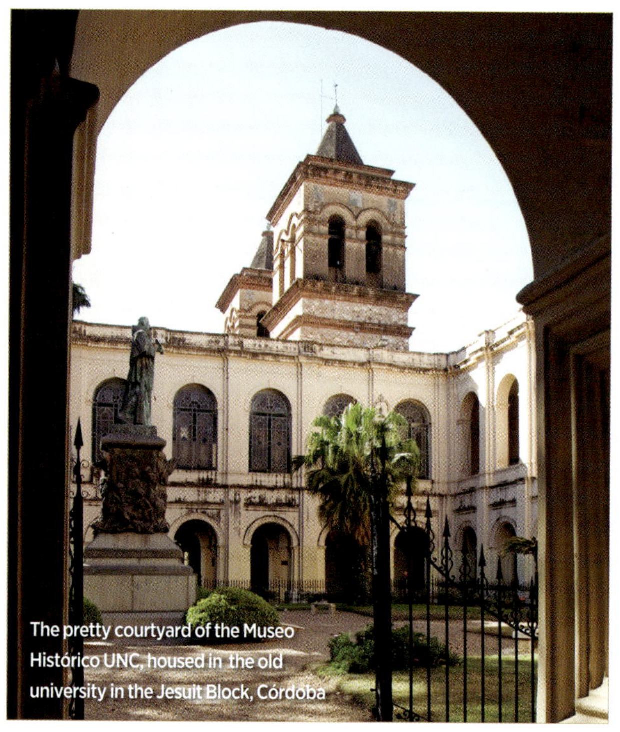

The pretty courtyard of the Museo Histórico UNC, housed in the old university in the Jesuit Block, Córdoba

percentages of Isabella and Viognier, while Bodega Del Gredal's *(delgredal.com)* small-production Misitorco Sauvignon Blanc from nearby San Pedro Norte is a herbaceous surprise among Córdoba's strong red pack. Try Del Gredal's vintages with superb Colonia Caroya salami, which has held a GI (geographical indication) since 2014, at La Cautiva Parrilla steakhouse in Jesús María *(@lacautivaparrilla)*.

History abounds at La Caroyense, which in its 1970s heyday picked millions of kilos of grapes. Today oenologist Agostina Lucchesi makes 500,000 litres of traditional-method espumoso sparkling, two grappa spirits and Lagrimilla, a sacramental wine, plus red varietals. A former co-op, the heritage is evident in oak casks and stained glass windows.

WHERE TO EAT Chef Martín Altamirano brings European Michelin experience to La Torgnole near Ascochinga, creating a whimsical seasonal tasting menu; sample Piensa's red blends here.

WHERE TO STAY Once the home of 19th-century president Julio Argentino Roca, rise and shine to sweet birdsong at Pueblo Estancia La Paz, a beautiful rural bolthole constructed in 1830 *(puebloestancialapaz.com)*.

Traditional Espiche (foam ceremony) is the first ceremony at Oktoberfest in Villa General Belgrano

A horseback rider explores the pampas trails through the Sierras Chicas hills near Córdoba

'A prevalent Germanic culture means Córdoba province hosts Latin America's largest Oktoberfest'

DAY 2 & 3

Valle de Traslasierra

There's plenty to savour in Traslasierra – a hilly southwest region about three hours' drive from Jésus Mária, dotted with charming towns and villages such as San Javier and Yacanto – from Granja Verbena's Sardo goat's cheese to the handcrafted local Fernet Beney amaro brand distilled with more than 40 highland herbs.

Overlooked by ruddy-faced Champaquí, Cordobá's tallest peak at 2,790m, Traslasierra offers activities galore such as cabalgatas (horse riding) with gaucho Alejandro Oliva of Los Teros, who also gives carriage-driving lessons, while anglers can fish for pejerrey (silverside) at Dique La Viña reservoir. Cooling off in one of the numerous shallow rocky streams is a treasured simple pleasure in summer. Winemaker Nicolás Jascalevich leads Traslasierra's movement at Bodega San Javier (*bodegasanjavier.com.ar*), setting out in 2001 to recover the region's lost wine heritage. Two decades on, standout bottlings that adhere to organic agriculture include Champaquí Gran Reserva Malbec-Cabernet Sauvignon and a stylish Noble Malbec rosé, while he also runs a cosy inn surrounded by vineyards.

Goyo and Ana Aráoz de Lamadrid also paired their winemaking project (in partnership with Richard Kirton) with a delightful lodge bolthole (*hotelbodega.com.au*) tucked away in San Javier. Besides cultivating 10ha of Malbec and Syrah vines with Mendoza-based viticulturist Federico Zaina, xerophile fan Goyo also keeps a 450-species Cactusarium and leads guided visits at weekends that culminate in a three-wine tasting with picada (charcuterie and finger-food).

Driving local sustainable identity is paramount at La Matilde, a biodynamic winery and farm plus intimate 10-room hotel run by Pablo Asef (*fincalamatilde.com.ar*). Viticulturist Matías Michelini helped Asef get the vineyard off the ground a decade ago, and although today Bodega San Javier's Jascalevich vinifies Malbec and Tannat here, Rhône-style whites are also on the horizon. The sustainable approach continues at De Adobe restaurant, with chefs picking organic vegetables from their own garden; it also stocks local Traslasierra wines, such as Finca El Boleado's Viognier and Bodega Viarago's Malbec duo; the latter opens its cellar door in Villa de las Rosas with prior reservation (*@viaragobodega*).

WHERE TO EAT Buenos Aires transplant Nitu Digilio left molecular cuisine (El Bulli in Catalonia, among others) to create abundant burgers at the charming 19th-century Peperina in La Población (*@peperinarestaurante*)

WHERE TO STAY Recharge batteries at Posada La Matilde, a kingdom of tranquillity and biodynamic vineyards (*posadalamatilde.com.ar*)

'Overlooked by Champaquí, Córdoba's tallest peak, Traslasierra offers activities galore'

DAY 4

Valle de Calamuchita

A three-hour drive east takes you on the winding mountain road known as the Camino de los Grandes Lagos, which makes for a radical contrast to the prairie-flat Colonia Caroya and Traslasierra's green hills. The principal Los Molinos dam, popular with water skiers and pejerrey anglers, guarantees a cooler climate in which the Calamuchita valley's white varieties are stating their case. While Oktoberfest may draw in the hop heads to Calamuchita's largest town Villa General Belgrano to celebrate, the Italian connection continues in Calamuchita, a relatively newer wine region. At Famiglia Furfaro, brothers Jorge and Hugo first planted on slopes overlooking Los Molinos in 2012, converting a potato farm into Villa Ciudad Parque's first vineyard (*famigliafurfaro.com*). Highlights from Famiglia Furfaro's mainly red portfolio include a powerful French oak-aged Cabernet Franc blend, while Primaterra Chardonnay's refreshing acidity is most enjoyable. Hugo runs a restaurant in Italy but he returns for harvest, making this the best time to visit, as the brothers are together and their infectious laughter reverberates around the log-constructed tasting room over a picada.

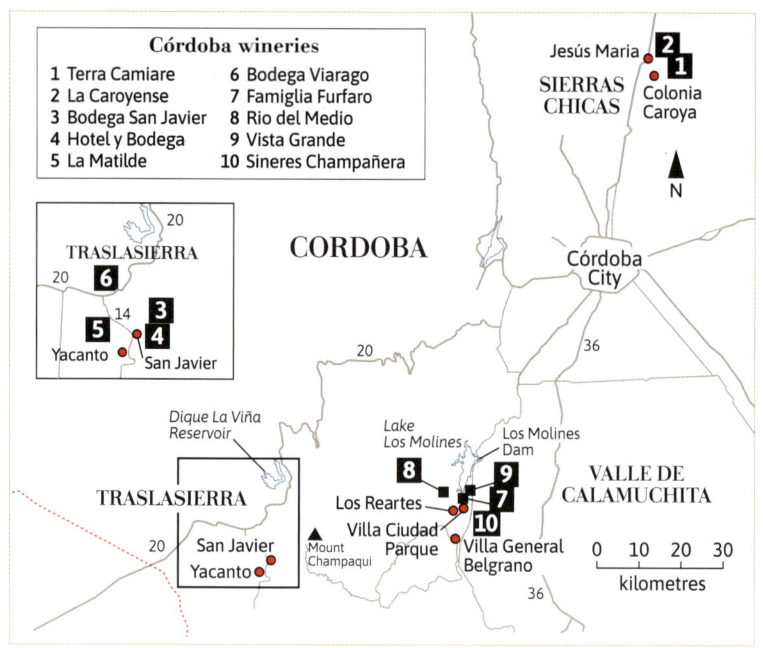

GETTING THERE

Córdoba is most easily reached by air, with numerous daily flights from Buenos Aires. Flight time is 1hr 23mins. Alternatively you can hire a car in Buenos Aires or take the bus; drive time is about 10 hours on RN9

Off the beaten track in Los Reartes is Río del Medio (*bodegariodelmedio.com.ar*), which is owned by Carlos and Laura Testa. The petit bodega's star is Malabar, a zingy Sauvignon Blanc named 'Argentina's Revelation 2021' in Decanter contributor Patricio Tapia's Descorchados guide, and it's best savoured while enjoying the rocky landscape and watery panorama of Los Molinos dam. Close by is Vista Grande (*fincavistagrande.com.ar*), an ideal spot to refuel with a picnic among the vines. Here, Daniela Martinelli leads this 4ha family project that started out as dad Daniel's hobby and today produces 14,000 bottles across nine labels. Her experimental approach is really paying off, co-fermenting Cabernet Franc and Merlot in stainless steel tanks and cultivating Rhône whites. Surmenage, a fresh Roussanne-Chardonnay blend, is particularly promising.

Los Molinos reservoir in Calamuchita

Sundown in Córdoba, where there are plenty of lively nightlife options

Calamuchita's latest bricks-and-mortar project is Sineres Champañera (@sineresespumante), which opened to visitors in October 2021. Husband-and-wife team Emiliano Guzmán and Andrea Fissore drive forward traditional-method sparkling wine, a radical take on bubbles for hop-loving Villa General Belgrano. For a Córdoba-wide panorama, Brazilian sommelier and regional transplant Cristiano Yamamoto, who used to lead the wine programme at Four Seasons hotel Buenos Aires, runs tastings in English, Portuguese and Japanese around Calamuchita *(mobile: +54 9 11 3014 9501)*.

WHERE TO EAT Beat the summer heat with cider on tap and a Neapolitan-style pizza at El Taller in Villa General Belgrano town, touted as the finest slice in all of Córdoba *(@eltallervgb)*.

WHERE TO STAY A stately hotel on the outskirts of Villa General Belgrano, Altos de Belgrano is a peaceful sanctuary away from the central hubbub (altosdebelgrano.com.ar).

DAY 5
Cordoba city

The provincial capital's dining and nightlife is vibrant and well priced, and a 90-minute drive back from Villa General Belgrano, so it's worth tagging on an extra night. After a day's sightseeing in the historic Jesuit Quarter or zipping alongside Suquía river on an e-scooter with Get Move *(@get.move.cba)*, refreshment is due with a pint of Golden Ale co-fermented with Isabella grape must at Höppers pub's spacious roof terrace *(@hoppers_cerveceria)*. Córdoba's nightlife is alive with pubs and cocktail bars in the Güemes district, such as Francis, and an evening is best rounded off listening (or dancing) to cuarteto at Estadio de Centro dance hall, fuelled by one last Fernandito.

WHERE TO EAT Devour the lunchtime tasting menu at El Papagayo *(elpapagayo.meitre.com)*, a slip of a restaurant artfully helmed by chef Javier Rodríguez; with a list of 30 or more Córdoba wines .

WHERE TO STAY A block from central Plaza San Martín, Azur Reál Hotel's creature comforts include a subterranean water circuit spa and Bruma restaurant *(azurrealhotel.com)*

Picking the grapes at Sineres Champañera

FACT FILE: MENDOZA

According to Wines of Argentina's 2021 annual report, Mendoza produced **76%** of all Argentinian wine across its **146,815ha**, cultivating some **39,250ha of Malbec** (20% of the national total). The other most widespread varieties included Bonarda (about 14,800ha), Cabernet Sauvignon (10,500ha) and the Criollas (12,500ha combined, mostly Criolla Grande).

'Maipú's quiet country roads lined with peach orchards, olive groves and vineyards are a breath of fresh air'

MENDOZA

Wineries of every size and style, breathtaking Andes views, mouthwatering fire-pit cooking, horseback adventures and Malbec, glorious Malbec – you'll find all this and more in Mendoza. The only question is: how much can you pack in?

There are said to be about 880 bodegas in Mendoza, ensuring that Argentina's elevated western province has long been a destination of choice for wine lovers. Vineyards here range from about 430m to 2,000m altitude, and while Malbec rules the roost, an ever-growing cast of varieties such as Semillon, Pinot Noir and the Criolla grapes (principally Torrontés, Criolla Chica, Criolla Grande and Cereza) means there's plenty for wine-curious travellers to savour beyond the headline-act red.

Whether it's horseback riding over the Andes or matching chocolate with wine, 300 days of sun and exciting wine-related activities keep visitors returning to key wine regions Maipú, Luján de Cuyo and Uco Valley. Late summer (early March) welcomes the arrival of the **Fiesta Nacional de la Vendimia** grape harvest festival *(vendimia.mendoza.gov.ar)*, while in winter (peak season July to September) powder lovers can hit Las Leñas' slopes, and après-ski on Malbec.

Languid paired tasting menus have long tempted foodies, but Mendoza's dining scene has rocketed over the past few years to claim the title of Argentina's most diverse food region (after capital Buenos Aires). Celebrity chef **Francis Mallmann** *(@francismallmann)* has long been associated with Mendoza, his open-fire techniques creating a show of their own. While the asado (barbecue) experience is guaranteed to please, a new wave of chefs is captivating palates putting wine first and showcasing star local products such as heirloom tomatoes and Andean native potatoes; veggie-led menus are finally in fashion.

A slew of restaurants has opened – and not just in bodegas. The glorious Andes range lends itself to outdoor dining experiences at restaurants such as **Cundo** *(@cundoaltamira)* and **Ruda** *(@ruda.cocina)* in Uco Valley, and Chirivia *(@_chirivia)* in Potrerillos. But 2023's most anticipated launch was **Angélica - Cocina Maestra** at Catena Zapata *(via 'Contact' on catenazapata.com)*.

There's more good news given that this surge in dining spots is being matched by hospitality. Recent openings include acclaimed winemaker Susana Balbo's **SB Winemaker's House & Spa Suites** in Chacras de Coria *(susanabalbohotels.com)* and **La Morada** in the Uco Valley *(lamoradalodge.com)*; these offer comfy accommodation to suit all budgets.

One of the global Great Wine Capitals network and the host in October 2022 of the World's Best Vineyards awards, Mendoza should be high on your list of must-visit wine destinations. With the five-day guide that follows, travellers can visit both traditional and contemporary bodegas while soaking up the ultimate in wine lifestyle...

DAY 1 Maipú

To the south of central Mendoza city, the eastern department of Maipú is where European varieties Cabernet Sauvignon and Malbec began to be cultivated in the mid-19th century, spurred on by a surge in immigration, particularly from Italy from the 1880s onwards: Maipú and Luján de Cuyo together are known as the Primera Zona ('first zone'). While it's usually considered that Mendoza produces mountain wines, Maipú is its lowest-elevated district, topping out at a relatively rather lowly 700m-940m above sea level.

While Maipú is often overlooked for being distant from downtown Mendoza, its quiet country roads lined with peach orchards, olive groves and vineyards are a breath of fresh air. Open farmlands mean wineries aren't rubbing elbows; it can take half an hour to drive between them, so hire a car (Mendoza's signage has come on significantly in the past three years), or even a driver, as taxis can be scarce. If you're staying in Mendoza city, you can hop on the **Metrotranvía tram**; alight at Gutiérrez station – just across the road is the well-located **Wine and Ride** *(wineandride.com.ar)*, where you can hire bikes and staff will help to plan a tour of local vineyards and wineries to suit your agenda. Another alternative is the hop-on, hop-off **BusVitivinícola** *(busvitivinicola.com)* which offers half- and full-day options.

Among its cluster of at least century-old bodegas, just a handful in Mendoza still use foudres of 40,000 litres or more – close to the tram station, **Bodegas López** *(bodegaslopez.com.ar)*, founded in 1898, is one of them. Fourth-generation winemaking director Carlos López and his brother Eduardo, general manager, respect tradition by continuing to create cask-aged Bordeaux-style reds, while driving forward with a line of daily drinkers including Sauvignon Blanc.

A top seller is Montchenot, a range of old-vine blends dominated by Cabernet Sauvignon, with bottlings cask-aged for five, 10, 15, 20 years or more. A free guided visit makes for a fascinating history lesson, while other tour and tasting packages are available, from about £5-£50; for an additional class, visit the **Museo del Vino y la Vendimia** museum a short distance away along Calle Ozamis. Other top Maipú tastings include **Bodega Trapiche** *(trapiche.com.ar)*, housed in its impressive 1912 Florentine-style building along Calle Nueva Mayorga, while **Luigi Bosca** *(luigibosca.com)*, further out along Ruta 60, recently converted its centenarian Finca El Paraíso property into a restaurant, headed by chef Pablo del Río. **Mil Suelos** *(@milsuelos)* is also due to open a parrilla experience – cooking large cuts of meat on a metal grate over an open wood fire – later this year.

Santa Julia's story *(santajulia.com.ar)* is more contemporary but no less legendary. When José Zuccardi developed a new irrigation system, he began cultivating vineyards in eastern Maipú district Santa Rosa in 1963 to showcase it. Taking up the baton from his father, José Alberto began to cultivate high-quality grape varieties such as Tempranillo in 1982, naming that project after his daughter.

Today, Julia Zuccardi is responsible for tourism and hospitality *(on the Spanish language site, click on the 'Turismo' tab)*, while her brother Sebastián is winemaking director at the certified organic winery. Aesthetes will enjoy browsing works by Mendoza-based artists, while the more active can tour the estate by bike. It counts two restaurants: enjoy a picnic or a full-blown asado experience at Casa del Visitante; its empanadas (filled pastry pockets) were named Argentina's best in 2018.

Meanwhile, Pan y Oliva uses extra virgin olive oil in every dish, the ideal vehicle to showcase Zuelo, youngest sibling Miguel's line of liquid gold.

WHERE TO EAT Choose from one of nine El Enemigo pairings to accompany a three-course lunch or dinner at **Casa Vigil** *(universovigil.com, @casavigil)*, close to the Mendoza river near El Paraíso.

WHERE TO STAY Well-appointed suites and log fires await at **Hotel Club Tapiz** *(club-tapiz.com.ar)*, a delightful former governor's home built in 1890.

DAY 2 Luján de Cuyo

The western side of Primera Zona, Luján de Cuyo (see map, p187) lends its name to one of two denominación de origen controlada (DOC) zones in Mendoza, while harbouring sub-districts including the Geographical Indications (GI) Agrelo, Las Compuertas and Vistalba. Its central hub is Chacras de Coria, former vineyards and farms gobbled up by private housing estates.

Regardless, Luján is home to a wide array of bodegas, such as vigneron **Carmelo Patti's** authentically rustic cellar door on San Martín, where you can sample Bordeaux-style blends *(search 'Carmelo Patti' on experiencemendoza.com for details)*, or the recently opened **Anaia Wines** *(anaiawines.com)* in Agrelo, about 20 minutes by road south on San Martín – here, in a tribute to the national tea drink, the producer has developed the first mate gourd-shaped concrete fermentation tanks.

Given the urban proximity, activities are located closer together, so those looking to burn off a few wine calories can pedal

Trapiche, to the southeast of central Mendoza

Planta Uno food hall in Godoy Cruz, Mendoza

to tastings. Elevation is very gradual, so there's no real need for a mountain bike; maps guiding you to the likes of Alta Vista, Clos de Chacras and Viamonte are provided when you rent from **Vistalba Bikes** *(vistalbabikes.com)* on Embalse Potrerillos.

Luján is home to a second Mendoza bodega that focuses on traditional winemaking. Though its building was constructed in 1890, the Weinert family takes pride in the art of cooperage, restoring old casks at **Weinert Bodegas y Cavas** *(bodegaweinert.com)*, which was founded in 1975. After visiting the red-brick cellars and beautifully crafted toneles casks, enjoy a vertical tasting that, if you're in luck with your timing, could even include a 1977 Malbec.

At the other end of the winemaking spectrum you'll find **Riccitelli Wines** *(matiasriccitelli.com)* in Las Compuertas on the western edge of Luján. Matías Riccitelli is both playful and serious, sourcing old-vine Chenin Blanc and Merlot from Río Negro, Patagonia, while also creating the low intervention and pét-nat range Kung Fu, which gets snapped up by the Buenos Aires hipster drinking set. Book a tasting and a table at the bistro, helmed by passionate seed collector **Juan Ventureyra**, whose plant-focused lunch menu is a refreshing break from asado.

Recent restaurant-in-bodega openings include chef Francis Mallmann's outdoor dining experience **Ramos Generales at Kaiken** *(kaikenwines.com)* off Roque Sáenz Peña; **La Jamonería at Vistalba**, nearby, further along the main road; and in the same area, new for 2023, enjoy charcuterie pairing at Mauricio Vegetti's **La Bodeguita at Lui Wines** *(@luiwines)*.

As many wineries only open for lunch, **Brindillas** restaurant *(brindillas.com)* in Chacras is great intel for a dinner date, while a 15-minute drive north in Godoy Cruz is **Planta Uno** *(plantaunomercado.com)*. Here, Bodega Lagarde's Sofía Pescarmona overhauled a former metals factory to create a well-curated indoor food hall that includes wine bars and small eateries, avoids big brand names and, importantly, opens until 1am.

WHERE TO EAT At **5 Suelos - Cocina de Finca** in Las Compuertas, chef Patricia Courtois recounts Argentina's history paired with the Durigutti brothers' wines on her 14-course Menú Historia *(durigutti.com)*.

WHERE TO STAY A welcome massage and in-room sauna is the first step to relaxing bliss at **SB Winemaker's House & Spa Suites**, Mendoza's most illustrious 2022 hotel opening that includes **La VidA** restaurant *(susanabalbohotels.com)*.

DAY 3 Agrelo

Cross Ruta 7 – the road linking the Atlantic ocean with the Pacific – to Agrelo, in the southern part of Luján de Cuyo. Home to a host of big-name wineries located at about 950m elevation, take your pick of fabulous vintages in a vineyard-to-glass situation, because the options are numerous and the wineries again in relatively close proximity, mostly off Ruta 7. You can opt to tuck into a paired lunch at **Ruca Malen** *(bodegarucamalen.com)*, a picnic at **Espacio Crios** *(susanabalbowines.com.ar)*, farm-picked pistachios at **Caelum** *(bodegacaelum.com.ar)*, a bubbly bistro lunch at **Chandon** *(chandon.com.ar)*, or a biodynamic tasting at **Chakana** *(chakanawines.com)*. Remember all bodegas require advance reservations.

Two Agrelo top guns are **Viña Cobos** *(vinacobos.com)* and **Catena Zapata** *(catenazapata.com)* – named World's Best Vineyard 2023 – located a 10-minute drive away. There are two carefully curated tastings at Paul Hobbs-run Cobos, now on its 25th vintage: and now Angélica - Cocina Maestra (see p186) gives it some architectural competition in the shape of a majestic Italian-style villa complete with watchtower and basement distillery. Throwing open its mighty glass and steel

La VidA at SB Winemaker's House & Spa Suites

doors in February 2023, the Catena family's many award-winning vintages take centre stage at its first venture into hospitality beyond tastings – fourth-generation managing director Laura Catena and winemaker Alejandro Vigil join forces to challenge chef Iván Azar to match dishes to wine.

Other recent Agrelo openings include **Quimera** *(@quimerabistro)*, just off Ruta 7, from **Achaval Ferrer**, whose culinary approach combines a farm-to-table concept with fire. If you're keen to squeeze in more tastings and eat on the hop, grab a home-cured ham sandwich at the food truck where Ruta 7 meets Cobos street. An unexpected experience is **Las Palapas** *(@laspalapas)*, housed next to vineyards and the scenic Potrerillos dam, a cool Sunday afternoon electronic music party that attracts internationally reputed DJs.

WHERE TO EAT Enjoy a hands-on dining experience picking your salad and crimping empanadas before savouring the lunchtime fine dining menu at **Zonda** *(lagarde.com.ar)*.

WHERE TO STAY Book a luxurious villa among vineyards then chill at the spa after a hard day's tasting at **Cavas Wine Lodge**, Mendoza's only Relais & Châteaux property *(cavaswinelodge.com)*.

DAY 4 Uco Valley

From Chacras de Coria, it's a 60- to 90-minute drive south to Tupungato, Tunuyán and San Carlos, the Uco Valley's three principal departments, where snow-capped mountains dominate the landscape. Seeing the 6,570m-high Tupungato volcano means Uco is within reach and a chance to get closer to nature – and the Andes. Grapes and orchard fruits have long been cultivated in the valley and **Salentein** *(bodegasalentein.com)* was a pioneer in making wine at high elevation in Tunuyán in the late 1990s; located on Ruta 89, the bodega boasts majestic architecture to complement its world-class Pinot Noir, Chardonnay and Malbec. Within the estate, stop by Killka gallery, whose exhibits bring together Argentinian and Dutch artists.

A five-minute drive from Salentein is **Domaine Bousquet** *(domainebousquet.com)*, a certified organic bodega that also produces kosher Malbec; ask to sample it on the one-hour guided visit. As sustainability is Domaine Bousquet's main philosophy, the seasonal menu at its restaurant, Gaia, is strictly organic, chef Adrián Baggio sourcing many ingredients from the winery's orchard.

In nearby Gualtallary, winemaker Matías Michelini's **Sitio La Estocada** *(sitiolaestocada.com)* is a lesson in biodynamic agriculture. Helped by grazing animals, Matías' family tend the vineyards from which he sources grapes for his Passionate Wine range. An experiential gem is the lunar cycle dinner, on full and new moon nights.
Half an hour's drive away on Ruta 94, **The Vines of Mendoza** *(vinesofmendoza.com)* has grown so exponentially since conceiving its private vineyard project – enabling members of the public to realise their dream of owning a vineyard and making premium-quality wine – it now houses successful spin-off winery projects spawned from the original idea, namely Corazón del Sol and SoloContigo. **SuperUco** *(superuco.com)*, the joint Michelini brothers project, is also based here; all three are open to the public. A recent addition to the Vines project is Mitre Fortín distillery, the producer (in a different location) of Principe de los Apostoles, Argentina's first premium gin brand.

A lovely wining and dining alternative is to mount a four-legged friend for a mountainous expedition. The gauchos from the **Cabalgatas de la Quebrada del Cóndor** *(@quebradadelcondor)* lead small groups through cattle pasture and into the Andes on relaxed steeds that might huff and puff on steeper slopes. The view from the top is incredible, proffering a true feel of the valley's breadth. Ride it or trek it – if you've worked up an appetite, asado (and a glass of vino) awaits at the log cabin.

WHERE TO EAT Breathe in fresh mountain air while clapping along to contagious live folk music, paired with empanadas, at the charming **Bodega La Azul** *(bodegalaazul.com)*

Relaxed tasting area at Viña Cobos in Agrelo

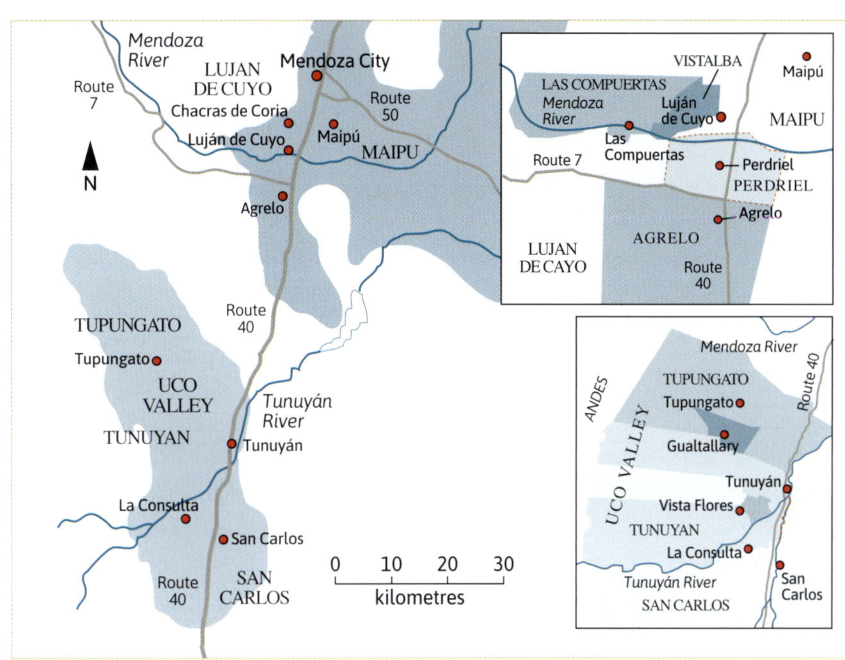

GETTING THERE

There are numerous daily flights from Aeroparque Jorge Newbery in Buenos Aires to Mendoza. Flight time is around 90 minutes. Hire a car at Mendoza's El Plumerillo airport; it's a 30-minute drive to Maipú and Luján de Cuyo, 90 minutes to the Uco Valley.

WHERE TO STAY Luxury hospitality pioneers in Uco, many of **The Vines of Mendoza's** spacious villas have fabulous Andean views; take your swimwear so you can enjoy a dip in the outdoor jacuzzi *(vinesofmendoza.com)*

DAY 5

San Carlos & Vista Flores

When a bodega picks up the World's Best Vineyard prize three consecutive times, you ought to squeeze everything you can from it. That accolade went to family-run winery **Zuccardi Valle de Uco** *(zuccardiwines.com)* in Paraje Altamira, San Carlos, between 2019 and 2021. It's fair to say third-generation winemaker Sebastián Zuccardi improves with age: his Finca Piedra Infinita Gravascal 2018 recently picked up 100 Parker points. Flavour-neutral concrete egg tanks let the vineyard parcels do the talking, while marvellous Andean views and succulent T-bone steaks are calling at **Piedra Infinita** restaurant; you can easily spend a day savouring a guided visit, tasting and paired lunch.

Another nearby estate where you can while away a day wining and dining is the Michel Rolland-founded **Clos de los Siete** *(closdelossiete.com)*, which houses the wineries of Flecha de los Andes, Monteviejo, Cuvelier Los Andes and Diamandes. An ambitious winery project, Clos de los Siete involves the four bodegas, run by four Bordeaux families and encompassing 850ha of vineyards, contributing elements to a single Bordeaux-style blend, while also making their own wines. And if you're in need of 360° panoramas, drop Gabriel Dvoskin of **Canopus Vinos** *(canopusvinos.com)* a line. His low-intervention, cool-climate Pintom Pinot Noir and Y La Nave Va Malbec from El Cepillo have been exciting Argentinian sommeliers for years; given that he hosts sporadic yet intimate vineyard tastings – conducted with as little embellishment as his vintages – it's worth getting in touch.

WHERE TO EAT Whatever the weather, wrap up warmly for an al fresco lunch at **Cundo** *(cundoaltamira.com.ar)*, not far from Zuccardi just outside La Consulta, where chef Seba Juez prepares a classy six-course Uco-focused menu.

WHERE TO STAY Check into a vineyard home at **La Morada Lodge** *(lamoradalodge.com)*, just west of Vista Flores, where the cellars are stocked by top sommelier Andrés Rosberg, then toast the peace of Los Chacayes.

MONTEVIDEO & CANELONES

If you're visiting South America, schedule a relaxed stop-off in Uruguay – a nation of 3.5 million people who love to dance and will offer a warm welcome. You'll discover the country's growing reputation for quality wines too.

Exploring Uruguay and its wine regions feels like you've just been let in on one of South America's best kept secrets. One of the smallest countries on the continent, Uruguay doesn't have the same bombastic personality as some of its Latin American neighbours, but sits as a silent siren for those in the know. Uruguay's steadily growing economy and progressive politics have made it a haven for international investment, and its sleepy capital city is increasingly cosmopolitan, with Uruguay's wine culture coming to the fore. As word gets out, there's no better time to discover its capital, Montevideo, and nearby wine route.

Discovering Montevideo

The tortured notes of the ivories being tickled are all the more soul-stirring under candlelight. The pianist expertly pulls us through undulating emotions as he pieces together tango songs that were first written on the streets of Montevideo a century ago. Although this tango dinner show at Primuseum (see p191) is number one on TripAdvisor, the small collection of warmly lit tables huddled around the piano and its pile of crusty old music sheets is satisfyingly intimate and personal. The friendly waiter pours me another glass of rich Tannat as I dig into my steak and wonder why Montevideo never received the same acclaim for its steak and tango as Buenos Aires.

Tango was, after all, invented between the ports and streets of both cities, and the steak is every bit as good (if not better, dare I say) in this country where cows outnumber people three to one. But Uruguayans don't boast about their claim to tango or steak. Nor do they often confess that they have the longest carnival in the world – their 40 days makes Rio's six look positively meagre. 'We don't really like to talk about ourselves too much,' a Uruguayan friend tells me the next evening over wine in a hip urban market, Mercado Ferrando (see p191). 'It just isn't our style.'

FACT FILE: URUQUAY

AREA PLANTED 6,343ha (26% Tannat)
WINERIES 176
EXPORTS TO 51 countries

'In Canelones, 90% of wineries are family-owned, and it's often the family who welcome you in'

Montevideo's Plaza Independencia, bordering downtown

GETTING THERE

Montevideo airport has daily flights to Madrid, Miami and Buenos Aires, or you can take a two-hour ferry from Buenos Aires.

Although no one will admit it, style seems effortless in Montevideo. The streets are a parade of architecture movements ranging from neoclassical giants like the Palacio Salvo and Teatro Solís theatre to belle-époque facades and modernist beach houses, which are all nonchalantly strung together. Even the airport has garnered design awards.

'Montevideo has more art-deco architecture than any city other than New York – and yet it's still off the radar as a destination,' British-born Karen Higgs, author of the Guru'Guay Guide to Montevideo, tells me over coffee in the Old City where she's been based since 2000. 'The secret delights of Montevideo are not immediately evident, which is what makes their discovery all the more delightful.'

Montevideo's streets can in fact feel eerily quiet during the afternoons, and it's hard to believe that one-third of the country lives here. In the world's most laid-back capital city, sipping yerba mate on the 22km seafront promenade constitutes a significant portion of weekend plans. In the evening, however, Montevideo is a hive of cultural activity – albeit mainly behind closed doors.

The Old City's historic bars and cafes are a good place to start, and hark back to the golden era of Uruguay's literati (including many tango composers). Catching a milonga dance is a quintessential Montevideo experience, but it is perhaps the murga that gives you a deeper insight into the idiosyncrasies of Uruguayan culture. This street performance combining political satire with comedy and song is a pillar of Uruguayan carnival, but performances and rehearsals are held year-round. Another rich cultural expression of Uruguay is candombe – an invigorating dance performed to the beat of many drums, which tells the tales of the African slave experience in Uruguay.

Canelones wine route

From culture to wine, the journey is easy, vineyards appearing before you reach the city limits – nearby Canelones became Uruguay's prime vine-growing territory in the 20th century precisely because of its proximity to the thirsty domestic market. The mild Atlantic climate is also conducive to quality grape production, with rich clay soils spread across the

Bodega Bouza has 7.5ha of vines near Pan de Azúcar hill in Maldonado

MY PERFECT DAY IN MONTEVIDEO & CANELONES

Reinaldo de Lucca

MORNING

Although locals drink yerba mate – an infusion of the dried leaves of a native holly plant, often drunk from a hollowed gourd – you may need coffee. **The Lab** *(thelab.com.uy)* has a great spot in Punta Carretas where you can start the day, coffee in hand, striding around the windy peninsula. Montevideo's greatest asset is being right on the water and a coastal walk is the most invigorating way to start (or end) the day.

After the morning traffic abates, head into Canelones where every winery visit is a personal affair. One of my favourite trips is to **Reinaldo de Lucca** *(deluccawines.com)*, a true vigneron in every sense of the word, who talks about vine-growing in the most poetic of ways.

LUNCH & AFTERNOON

For lunch, head to Lo de Porro (Batlle y Ordóñez 664) in Las Piedras, a typical bar of yesteryear where wine is served by the jug and pasta is freshly rolled each day. Afterwards, it's an easy drive to the **Pisano brothers** *(pisanowines.com)*, 10 minutes back up the road in Progreso. Each of the brothers (Daniel, Gustavo and Eduardo) has unique insights into Uruguayan wine (sales, winemaking and viticulture respectively) and a tasting with them is unforgettable. Ask to taste with the youngest generation of Pisano, Gabriel, who is at the forefront of Uruguay's new-wave wines with his boutique label **Viña Progreso** *(vinaprogreso.com)*.

Pisano Wines

Nightlife in the Old City

EVENING

Nightlife starts late in Montevideo, so take an early-evening stroll around the Old City absorbing the ambience, street art and architecture. Then spend the rest of the evening nibbling and imbibing while hopping between the growing number of wine bars and cocktail spots in the city, before ending with some live music or a milonga, which continue into the wee hours.

undulating hillsides which channel refreshing coastal breezes – essential in this more humid climate.

Although Canelones hosts two-thirds of Uruguay's wine production, 90% of the wineries are family-owned and it is often the family who welcome you in. Most are boutique producers, and each family puts its own unique stamp on its wines – as a result, exploring Canelones provides a wealth of diversity in wine styles and varieties.

'A big difference in Uruguay [compared to Chile and Argentina] is that we do experience significant vintage variation here, which keeps us on our toes!' explains Eduardo Boido, winemaker at Bouza *(bodegabouza.com)*, which sits at the gateway of Canelones. 'Some years are better for white varieties and others for red, but Tannat emerged as Uruguay's champion because we get great colour, acidity and concentration year on year.'

Tannat is Uruguay's most widely planted grape variety, but there are many others that show promise, including Albarino. The Bouza family was the first to plant this Galician white grape, which thrives in Uruguay's similar Atlantic conditions, as an ode to its Galician ancestors. This Spanish flair also makes its way onto the menu at Bouza's excellent restaurant, which vies for attention with its extensive vintage car collection.

Another top spot for lunch is Artesana *(artesanawinery.com)*, some 30 minutes' drive deeper into Canelones. This boutique

The barrel ageing cellar at the Bodega Juanicó Familia Deicas winery

winery was the first to plant Zinfandel, inspired by the California-based owners, and its outdoor restaurant among the vines is an excellent place to sample Uruguay's only Zinfandel paired with a wood-fire menu.

The Pizzorno family *(pizzornowines.com)* also offers an intimate lunch and tasting, where you can explore its 80-year winemaking heritage and allow your mind – and tannic preconceptions – to be blown by tasting Uruguay's first carbonic-maceration Tannat.

Another interesting exploration of Tannat is tasting the Familia Deicas terroir range at Juanicó *(juanico.com)*, one of Uruguay's leading producers with the oldest cellar in the country, constructed in 1830. Other notable historic wine families to visit include Carrau *(bodegascarrau.com)*, Antigua Bodega Stagnari *(antiguabodegastagnari.com.uy)*, Varela Zarranz *(varelazarranz.com)* and Los Nadies *(bodegalosnadies.com)*, ranging from major players to boutique.

There's no lack of cellars to discover tucked into the folds of Canelones and Montevideo, and the wine families of this region will encourage you to continue your discovery of Uruguayan wine by visiting the nearby wine routes of Atlántida, Colonia and Maldonado too. Start planning your next trip to Uruguay now – you've just been made privy to South America's best-kept wine secret.

YOUR MONTEVIDEO & CANELONES ADDRESS BOOK

ACCOMMODATION

Casa Sarandi For a home away from home, Casa Sarandi B&B offers plenty of character, comfort and all the insider information you could want. A cultural immersion in Montevideo's Old City. **casasarandi.com**

Sofitel Montevideo This 1921 art deco hotel is dubbed 'palace in the sand' for its prime beachside location in upmarket Carrasco. The epitome of opulent luxury, with handsome suites, a great restaurant, well-stocked cellar and a ritzy casino. **sofitel.accorhotels.com**

RESTAURANTS

Alquimista Tucked away in a peaceful corner of Carrasco, this B&B-turned-restaurant has tables set in different rooms of the house and garden, making you feel more like a guest than a diner. The innovative and colourful Uruguayan dishes are top restaurant quality. **alquimistamontevideo.com**

Mercado del Puerto Eating at Montevideo's main market is more about the all-round experience than the quality. A carnivore's delight, your eyes will water at the sight of so much asado (slow cooked barbecue) – and that's before the smoke hits. **mercadodelpuerto.com**

Primuseum If you want a side of tango with your steak, Primuseum is the place for you. This intimate restaurant set in an antiques museum in the Old City serves a

Barolo

Uruguayan tasting menu while local musicians deliver a captivating show. **primuseum.com**

WINE BARS & SHOPS

Barolo The impressive cellar of Barolo stocks some 160 labels which can be ordered by the glass or flight, or uncorked at Fellini restaurant next door. **fellini.uy**

Madirán & Mercado Ferrando
This urban market has several eateries, bars and boutiques ranging from gastronomy book shops to artisanal tap houses. Wine lovers should visit Madirán wine bar for its eclectic selection. **mercadoferrando.com**

Montevideo Wine Experience
Under the expert eye (and fluent English conversation) of Nicolás and Liber, a couple of hours here will give you a whirlwind introduction to Uruguayan wine. Stay late for the live music sessions. **See Facebook**

Mercado del Puerto

GEORGIA

After decades of conflict that obscured much of its rich cultural heritage, this country is rediscovering itself. Celebrate its rebirth with a winding adventure from the bars and restaurants of the capital to the family-run wineries of rural villages.

'Being there for the first time, I felt I'd finally found home'

Three of my grandparents were born in Italy; they emigrated to America before World War I but never forgot their Italian childhoods. My Piemontese grandmother, Marie, often talked to me about her own grandmother. She lived simply in the country, had a pet pig called Cleopatra, made cheese from her cows' milk, kept silkworms, extracted lanolin from sheep's wool and grew vegetables and grapes for the family's wine.

I was resident in Italy for more than 20 years and spent much of that time living in – and writing about – rural communities in several regions. By then, European Union rules had ended this type of integrated agriculture. Even in areas where monoculture was shunned, it was no longer possible nor desirable to produce a little of everything. Vineyards are rarely still interspersed with fruit trees; you can no longer keep a few goats or cows for home use: most animals have been grouped indoors with more industrial husbandry. I always regretted not having been able to experience Sicily or Piedmont as it was then. Self-sufficient agriculture appeals to me and seems ever more important.

Maybe that's why I fell in love so quickly with Georgia. Within days of being there for the first time, eight years ago, I felt I'd finally found home. Driving west from Tbilisi with a group of fellow wine lovers who, like me, had attended the second

'From polyphonic song and dances to regional recipes and native grapes, this is an exciting time of rebirth in Georgia'

international qvevri symposium, I parsed the landscape from my window. Here were simple, two-storey houses surrounded by vegetable patches, fruit trees and vines that resembled my childhood stories. Fields were small and often flanked by woods or decorative wrought iron. As our coach slowed and wove around cows idling in the road, or passed pigs lazing in muddy ditches, I felt a kind of thrill.

So many animals are free in Georgia, they're a symbol of the country's desire for self-determination. (The animals have owners, but they're let out of their pens each morning to spend the days as they choose.) I might not yet have mastered Georgian, but I understood something of the country's rural lifestyle.

Slow the pace

Georgia is undergoing a period of self-discovery after the fall of the Soviet system, of which it was an unwilling part. If the Soviet times imposed a vision of standardised industrialisation and tried to cancel much of Georgia's cultural heritage, today the focus is on unearthing and celebrating those native customs. From polyphonic song and dances to regional recipes and native grapes, this is an exciting time of rebirth in Georgia.

The best way to experience these things is to visit rural winemakers. Many hospitable families have begun receiving guests in their homes and wineries (often the same thing), offering meals, wine tastings, music and more. Of the dozen or so regions in Georgia, it's primarily the central ones that produce the most wine. (The mountains of Svaneti, Kazbegi and Tusheti in the north are too high for vines to grow.) While eastern Kakheti is still by far the largest producer of grapes, the central and western parts of the country are full of fascinating people working with the ever-expanding range of native Georgian varieties.

My advice for those wanting to travel in Georgia is to do it slowly. Spend a few days in Tbilisi and then take the time to wander through small villages and side roads to really experience the countryside. The easiest way is to hire a driver (usually also the most economical solution) who can get you to what are often obscure villages and wineries. Road signage has recently improved, but navigating is sometimes tricky.

While many producers may not have modern websites, they are all on Facebook; that's the best place to reach them. Always make appointments before heading into the countryside as producers are not always available. Here is a short compendium of places to go in central and western Georgia, by region. There are many more!

Tbilisi

This beautiful city is fun to explore on foot. From the historic old town, built onto the steep Mtkvari river bank, wander up little streets of colourfully painted wooden houses – many of which are now hotels – to Narikala fortress above. From there take the funicular for a bird's-eye view of the Peace Bridge and the presidential palace. The main street, Rustaveli Avenue, is home to many grand buildings, including the

National Museum, which houses ancient qvevri from digs, and fine early gold artefacts *(museum.ge)*.

For a lunch of home-cooked flavours, try the tiny Salobie Bia in Ivane Machabeli Street, if you can get in. Or, in warm weather, go across the street to the Writer's House for a more elegant meal in the villa's shady gardens. I love eating at Vino Underground, where the cooked-to-order food pairs so well with the wines. This natural wine bar is still the hub for Georgia's family growers, and it's the best place to sample or buy qvevri wines *(Galaktion Tabidze Street)*.

If you like this neighbourhood and are on a tight budget, check out Black Tomato Hostel *(see Facebook)* – it's very cool! Higher up, with great views over the city, Hotel Gomi 19 is also affordable and incredibly welcoming *(hotelgomi19@gmail.com)*.

In the last five years, lots of great new places have opened. Stamba is a stunningly designed central hotel, brasserie and garden in an ex-printing factory. It's a sister project to the trendsetting Rooms Hotels and, between them, they have revitalised the area around Rustaveli Metro *(stambahotel.com)*. Alubali is a fun, relaxed place for a drink or bite, in the courtyard garden, and it's open 'til late *(Akhvlediani Street)*.

The Wine Factory is another industrial conversion featuring several bars and restaurants, including the latest version of chef Tekuna Gachechiladze's Culinarium cooking school and chef's table *(see Facebook)*. A few streets away, at Saidanaa

Ancient vine terraces in the Mtkvari river valley

(From left) Zaza Gagua and Keto Ninidze with Nika Partsvania of Vino M'artville

Archil Guniava's qvevri cellar

Lunch at Oda Family Marani

Scientific Research Centre of Agriculture

wine – of local white Chinuri grapes – are exceptional here *(chardakhi@gmail.com)*.

Don't miss the iconic Svetitskhoveli cathedral in Mtskheta, once the country's capital and a Unesco World Heritage Site. And just outside Mtskheta, on the Tbilisi bypass road, Salobie is Georgia's favourite fast-food restaurant *(see Facebook)*, selling delicious stewed beans with cornbread, kebabs and juicy khinkali dumplings.

Imereti

Further west from Tbilisi, Imereti is a region of small hills, farms and woodland, and is home to many fine family wine producers. They work in qvevri as this is a production centre for the clay winemaking pots. At Maqatubani, on the main road from Khashuri to Kutaisi, Zaliko Bozhadze *(qvevri.maqatubani@yahoo.com)* is a master potter whose wares can be seen from the road. His studio and large kiln are a few steps down from it.

Archil Guniava, in the tiny village of Kvaliti, is one of the region's finest winemakers *(archilguniavawinecellar@gmail.com, or see Facebook)*. Don't miss his qvevri cellar, with buried vessels of many sizes, it's my favourite! His wines of white Tsolikouri and red Otskhanuri Sapere are fresh and drinkable.

Closer to Kutaisi, at the unpronounceable Nakhshirghele, the enfant terrible of Georgian wines, Ramaz Nikoladze, makes some of the purest, most complex of the qvevri wines of white (or amber) Tsolikouri. His wife Nestan is a fabulous cook too *(georgianslowfood@yahoo.com)*. Kutaisi, the Imereti region's capital, has a great covered food market. Before the coronavirus shutdown, airlines flew into its Kutaisi airport direct from European cities. Hopefully they will again.

Samegrelo

In the country's far west, the Samegrelo region borders the Black Sea and has a more temperate, humid climate than eastern Georgia. One of my favourite places to eat and taste wines here is at Oda Family Marani, outside Martvili, which is famous for its spectacular river caves. It's run by Keto Ninidze

Ramaz Nikoladze

Iago Bitarishvili's wines

Khachapuri

(@saidanaa_), Nathan Moss is producing high-quality charcuterie from Georgian pigs, and is opening a café to feature his products (Sharadhidze Street). They go so well with the qvevri wines.

Kartli & lower Mtskheta Mtianeti

This central part of Georgia is due west and a little north of Tbilisi and can be reached in a day trip from the capital. To see Georgia's 425 or more native grape varieties, book to visit the Scientific Research Centre of Agriculture in Saguramo *(srca.gov.ge)*, an impressive viticultural institute. The national collection, begun in 2009, extends over 44ha of vines – a row for each variety. You can't taste the wines these grapes make, but it's a fascinating place, especially in season when the grapes are visible.

Not far away is Chardakhi village, where pioneering qvevri winemaker Iago Bitarishvili and his wife Marina Kurtanidze *(iago.ge)* have their cellar and home restaurant. The food and

'Many hospitable families have begun receiving guests in their homes and wineries, offering meals, wine tastings and more'

and Zaza Gagua, who each produce characterful wines from rare local varieties including the light red Ojaleshi, Orbeluri and Dzelshavi *(Gagua under the banner of Vino M'artville with his friend and partner Nika Partsvania)*. Keto has built a fascinating outdoor wicker room with a dirt floor and open fire on which she cooks many traditional dishes *(oda.wines@gmail.com)*.

If you're driving to or from the highlands of Svaneti, stop for a meal in Zugdidi, at Diaroni *(diaroni.ge)*. This large restaurant serves local fare, from cheesy cornbread to spicy ribs.

Guria & Adjara

South of Samegrelo, these regions lead down the Black Sea coast to the picturesque seaport of Batumi. The vegetation of citrus and exotic plants is unique in Georgia. Until recently, tea was cultivated here and exported throughout the Soviet Union. Hazelnuts are now the favoured cash crop, but a few growers are working to re-establish the coast's native grape varieties, especially Chkhaveri. Zurab Topuridze, with his Iberieli brand wines *(iberieli.com)*, was the first to bottle it, but other small producers now follow his lead.

Chkhaveri produces a super-drinkable, light and refreshing red-rosé that goes well with the seafood from the coast, especially the noble Black Sea turbot. You'll find it at the daily fish market in Batumi, and can get it cooked for lunch next door at the small restaurant Balagan Fish and Grill on Gogebashvili Street *(gobatumi.com)*. Don't miss the other local speciality: egg-topped, cheesy khachapuri bread shaped like a boat.

Kakheti

For more places to visit in eastern Georgia and Tbilisi, see my Travel feature in Decanter June 2016 issue, also available on Decanter.com ('Georgia: Restaurants, hotels and shops').

Batumi's astronomical clock tower

Svetitskhoveli cathedral

WORLD'S 50 BEST WINE TRIPS

Decanter has been sharing expert travel advice for decades. Discover the 50 wine trips and experiences most read about on *Decanter.com* and you'll have enough ideas to fill years of travel. Explore Burgundy by bike, stay in a Bordeaux château, soak up the atmosphere in Jerez's Sherry bars, and visit South Africa's top wine hotels – it's all here, with dozens of other inspiring ideas. Full articles can be found at **www.decanter.com/top-50-travel-2023/**

Chapel Down Vineyard, Tenterden

THE WINERIES OF KENT

Southeasternmost county Kent is often described as the Garden of England. Stretching from the Thames estuary to the English Channel, you'll find a verdant landscape of rolling hills, blossom-filled orchards and white-cowled oast houses. It gets more sunshine and higher temperatures than most of the UK, which explains why it's famous for fruit. In places, you would be forgiven for thinking you were in France's Champagne region – you don't have to divert far to find patches of vines shimmering in the breeze on south-facing chalky slopes. Output is growing, with warm summers spelling bumper crops, and demand high. Little wonder Kent winemakers – namely Biddenden, Chapel Down, Domaine Evremond, Gusbourne, Hush Heath, Simpsons, Squerryes and Westwell – have redubbed this the Wine Garden of England (*winegardenofengland.co.uk*).

Castelvetro, Modena

LAMBRUSCO LANDSCAPES

With under-explored cities of art, supreme sports cars and glorious countryside, a trip through the land of Lambrusco is a must. Today's Lambruscos are shedding the stigma of cheap-and-cheerful predecessors thanks to a new wave of small, independent wineries. In Modena (Emilia Romagna) – home to Ferrari and balsamic vinegar – Lambrusco is present everywhere. There is a huge sculpture of grapes made of Murano glass (on the main roundabout along the Via Vignolese towards Vignola), as well as restaurants and bars including Lambruscheria, on Calle di Luca in the city (*lambruscheriamodena.it*). North of Modena, pale Lambrusco di Sorbara is traditionally made as a field blend of the Sorbara grape and Lambrusco Salamino. Across the regional border in Lombardy, Franco Accorsi at Fondo Bozzole named his Giano wine (£18 Villeneuve Wines), a 100% Lambrusco Salamino with notes of black cherry and chocolate, after Janus, the god of new beginnings.

48

Erbusco

FRANCIACORTA'S FINE VIEWS

Blending rural charm and tradition with vineyard-lined hills, Franciacorta is prime sparkling wine territory in Lombardy, northern Italy. Any tour should include a visit to wine town Erbusco, between Brescia and Bergamo, where Vittorio Moretti – a leading player and former president of the Franciacorta consorzio – is a reference for the area at Bellavista *(bellavistawine.it)*. This first-class winery is named after the location of its vineyards on top of the Bellavista hill. Besides the splendid views, the blend of art and sculpture (plus a waterfall and swing) is enchanting.

PIEDMONT: TRUFFLES, CASTLES – AND WINE

At the foot of the Alps in northwestern Italy, Piedmont's landscapes look like a living painting. The region is famous for its 'Three Bs' – Barolo, Barbaresco and Barbera. The first two, made from Nebbiolo, are the most prestigious wines. UNESCO-listed Langhe e Roero, with its bucolic landscape dominated by vineyard hills, offers white truffles as well as wine; Alba town holds a truffle festival each autumn. Wander among the castles, through the villages and museums (particularly the WiMu wine museum and Corkscrew Museum, both in the village of Barolo), and take in the panoramic views over the Langhe. Another World Heritage Site, Monferrato is famous for its hand-dug cellars, and is home to Barbera and sparkling Asti.

47

Autumnal hills and vineyards surrounding Barolo village

46

La Rioja Bike Tours

CYCLING IN RIOJA

Rioja is an ideal destination for any keen cyclist. With scenic views flanked by the Cantabrian mountains, rural roads and paths negotiable by bike, and providers such as La Rioja Bike Tours *(lariojabiketours.com)* happy to deliver to your accommodation – hybrid bikes or e-bikes are recommended – it's the perfect way to take in many of the most famous names in Spanish wine and the sea of vines that surround them.

Logroño, the capital of Rioja, makes an ideal starting point. It has the charm of any historic Spanish city: labyrinthine alleys leading to vast open plazas, an enchanting mix of Romanesque and Gothic architecture, plane trees reaching to the skies. However, there is one street in Logroño that makes any visit worthwhile. As evening comes, Calle del Laurel erupts into life. On the stroke of eight, locals of all generations burst onto this alley, for this is when the pincho (like tapas) bars open. They aren't designed for lingering; enjoy your pincho and small drink and move on to the next. It's a sociable, gastronomic pleasure and despite the spectrum of local wines on offer, 99% of locals drink Rioja crianza. There is a mild adjustment as you get used to the red wine sitting on ice (to combat the ever-present heat), but this is about living like a local, and you quickly get used to it.

A WEEKEND IN THE JURA

Neighbouring Burgundy may seem the more obvious travel choice, but Jura has its own delicious flavour. Driving is your best bet for exploration. Head straight for Arbois village to intimate chambre d'hôtes Closerie les Capucines, refurbished by importer Neal Rosenthal. Evening aperitifs await at Le Bistrot des Claquets, a no-frills wine bar (a favourite with local producers), and dinner at Le Bistronôme *(le-bistronome-arbois.com)*. Next morning head southwest to the scenic Château-Chalon or L'Étoile area for a breathtaking drive to local wineries such as Domaine de Montbourgeau or Domaine Berthet-Bondet. Later, visit forward-thinking Valentin Morel or tour the Musée de la Vigne et du Vin du Jura *(arbois.com)* back in Arbois.

VERDE VALLEY, ARIZONA: NAMES TO KNOW

The Grand Canyon may always be Arizona's biggest drawcard, but among the jagged red-rock formations, the state's newest AVA (American Viticultural Area), Verde Valley, is turning heads, too. Arizona wine dates back to the 16th century when Spanish missionaries planted grape vines, but the advent of Prohibition in the early 20th century halted production. The first official AVAs included Sonoita and Willcox, near the Mexican border, but the more recently recognised Verde Valley *(verdevalleyava.org)* is two hours' drive north of Phoenix. This used to be a sleepy part of the state, overlooked by travellers on their way to Sedona, but the wine industry's growth has led to a significant uptick in tourism. Look out for names such as Caduceus Cellars, producer of Monastrell and Tempranillo, and Chateau Tumbleweed, bottling Rhône and Italian varieties. Page Springs Cellars, from trailblazer Eric Glomski, Javelina Leap Vineyard & Winery and DA Ranch should be on your list, too.

Chateau Tumbleweed

TOP HEALDSBURG WINERIES

At the centre of three diverse Sonoma County wine regions – Russian River Valley, Dry Creek Valley and Alexander Valley – Healdsburg has it all. Cool-climate Pinot Noir and Chardonnay, old-vine Zinfandel, and world-class Cabernet Sauvignon. For a taste of Bordeaux and Burgundy in Alexander Valley, head straight to Jordan Winery's ivy-clad, French-inspired chateau *(jordanwinery.com)*. Cabernet Sauvignon and Chardonnay are made here in an Old World style focusing on approachability and food-friendliness. The 485ha estate is known for its picnics, vineyard hikes and food pairings made with ingredients sourced from the estate gardens.

At the other end of the spectrum, J Vineyards *(jwine.com)* in Russian River Valley has built its reputation on traditional-method, cool-climate sparklings. The J Bubble Room hosts five-course food and wine pairings on weekends. Finally, Dry Creek Valley estate Ferrari-Carano *(ferrari-carano.com)* is one of the most beautiful in California. Bottlings run the gamut from Italian varieties to sweet dessert wines. But the main event here is the gardens. Meander through 2ha of Italian and French-style geometric hedges and grand fountains. Every spring a flood of colour stems from 10,000 tulips and daffodils blooming.

Ferrari-Carano Vineyards & Winery

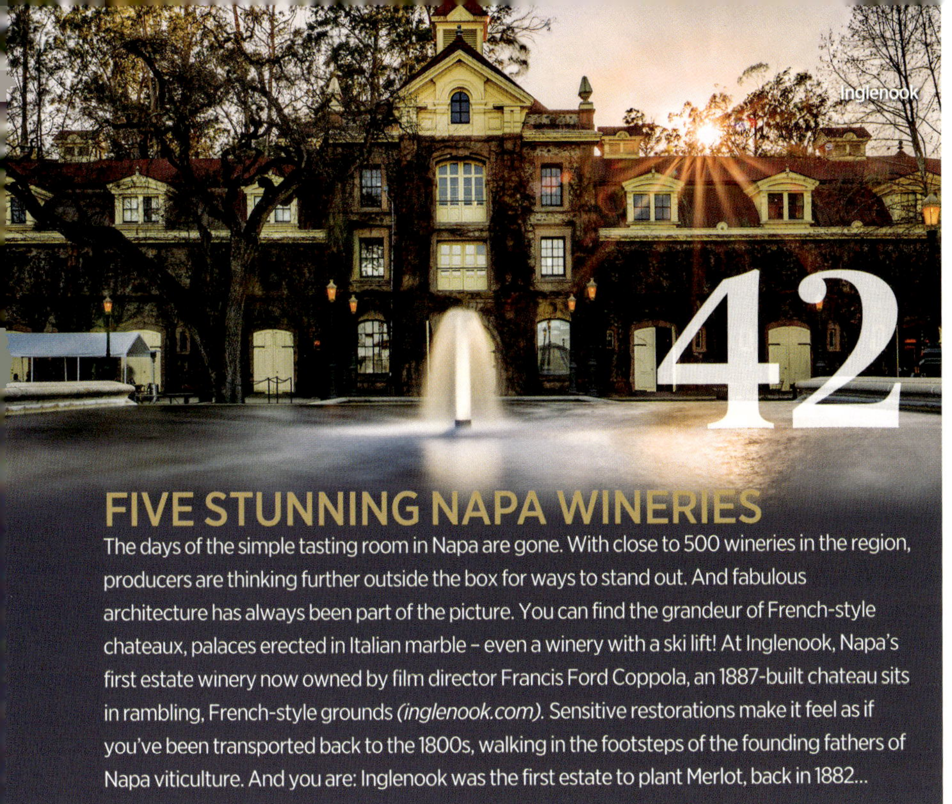

Inglenook

FIVE STUNNING NAPA WINERIES

The days of the simple tasting room in Napa are gone. With close to 500 wineries in the region, producers are thinking further outside the box for ways to stand out. And fabulous architecture has always been part of the picture. You can find the grandeur of French-style chateaux, palaces erected in Italian marble – even a winery with a ski lift! At Inglenook, Napa's first estate winery now owned by film director Francis Ford Coppola, an 1887-built chateau sits in rambling, French-style grounds *(inglenook.com)*. Sensitive restorations make it feel as if you've been transported back to the 1800s, walking in the footsteps of the founding fathers of Napa viticulture. And you are: Inglenook was the first estate to plant Merlot, back in 1882...

Marc Haeberlin

ALSACE: BEST BARS AND RESTAURANTS

What makes a visit to this northeastern French region especially appealing is its culinary scene. Brasserie Aux Armes de Strasbourg *(auxarmesdestrasbourg.com)* near the city's cathedral provides a jovial, pleasingly kitsch setting: wooden furnishings and red and white tablecloths. Opened in 1900, the food remains classic Alsatian, from baeckeoffe (traditional casserole) and choucroute, to chicken suprême with Riesling. For a more upscale experience, over on the south side of the river Ill, Les Haras Hôtel is set in a former royal stable, and the Brasserie restaurant *(les-haras-brasserie.com)* features both traditional fare and creative dishes with menus supervised by Marc Haeberlin, chef patron of the Michelin two-star Auberge de l'Ill in Illhaeusern. As for wine bars, L'Alsace à Boire *(alsaceaboire.fr)* lists examples of nearly all of Alsace's 51 grands crus, with by-the-glass selections to go with cheese plates.

Château St Martin de la Garrigue

TOP LANGUEDOC ESTATES

Wine tourism in the southerly French region Languedoc was relatively undeveloped until recently, but things are gradually changing. One of the most famous wines here is Picpoul de Pinet – the firm, salty dry white that accompanies the local oysters so well. Château St Martin de la Garrigue outside the village of Montagnac produces some of the best Picpouls *(stmartingarrigue.com)*. Meanwhile, for Syrah, Domaine de l'Hortus (domaine-hortus.fr) is one of the pioneering estates of the Pic St-Loup appellation. For a wine tasting almost within the city boundary of Montpellier, Château de Flaugergues – a 17th-century château with elegant gardens *(flaugergues.com)* – offers a range of wines covering Grès de Montpellier and Méjanelle.

VIENNA'S URBAN VINEYARDS

A long weekend in a wine region is about more than discovering new wines. You get a sense of the place, its people and their traditions. Take Vienna. It may not be an obvious destination for a wine tour, but it's the world's biggest urban vineyard region, with 276 producers working about 600ha – all within sight of the Stephansdom cathedral. Wine is an intrinsic part of the Viennese way of life; every weekend locals hang out in heurigen (producer-run wine taverns) or in buschenschank (open-air bars) among the vines. At Wieninger *(wieninger.at)*, on the sloping, south-facing Bisamberg – just a 20-minute drive from Stephansdom – owner Fritz Wieninger is credited with putting gemischter satz (field blend) wines on the map as Vienna's USP.

Wieninger

BEST SHERRY BARS IN JEREZ

The Sherry trade made the small city of Jerez an economic powerhouse, reflected in its grand architecture. Jerez today is quite different, though. It has resisted globalisation, favouring the quiet life. Things move slowly; people stop to talk, to have a drink. In tabanco bars locals originally bought tobacco and wine in bulk to take home. Now, friendly residents stand cheek by jowl year-round, sipping cheap and excellent dry wine, often watching spirited flamenco performances. Traditional tabancos – such as Tabanco El Pasaje *(tabancoelpasaje.com)* – still serve Sherry directly from the barrel. More and more establishments, however, are putting food first, such as Albores just off Plaza del Arenal *(restaurantealbores.com)*, serving Andalucía's prized almadraba tuna.

NAPA ACCOMMODATION FOR EVERY BUDGET

Napa Valley has an unrivalled choice of places to stay. St Helena makes a fine base for exploring the many vineyards, and here a relative newcomer to the luxury resort scene, Las Alcobas *(since rebranded as Alila Napa Valley: hyatt.com)*, has become a go-to stay for those seeking vineyard views. Situated in an old Georgian house, it feels more like a private estate than a hotel. Historic Beringer Vineyards *(beringer.com)* sits adjacent and many suites look out onto those very special vines. Dine on your own private terrace, or try the modern California cuisine at the hotel's Acacia House. There's an outdoor pool and fire pit, too, and the concierge can help organise winery visits, including transport – which may or may not include a hot air balloon ride over the vines...

VALLE D'AOSTA FOR WINE LOVERS

Bordering France and Switzerland, the tiny northern Italian region of Valle d'Aosta boasts famous Alpine peaks – and a cultural duality that charmingly fuses Italian and Gallic sensibilities. This small and sparsely populated region is almost too mountainous for agriculture, yet has an astonishing array of grapes. Among soaring peaks and fairytale castles, find French favourites Chardonnay, Pinot Noir and Gamay, as well as high-altitude Nebbiolo and Pinot Grigio (here called Nus Malvoisie). The city of Aosta is a solid home base from which to discover small, family-run wine estates such as Ermes Pavese *(ermespavese.it)*, which produces wines from the indigenous Prié Blanc grape. Wines capture Alpine purity, invitingly fresh with crisp acidity, and the family's homemade crisps (potato chips) make for a memorable tasting in the cosy cellar.

Tain l'Hermitage

EXPLORING THE NORTHERN RHÔNE

It used to amaze me that the home of Hermitage was so uninterested in tourism. Fifteen years ago, I arrived late into the village of Tain l'Hermitage and, though it was only 9.30pm, I couldn't find anywhere open for dinner. A lot has changed since then. In this unspoilt landscape of steep, terraced vineyards – some of the most dramatic in the world – there is a lot to enjoy. Squeezed in between the river and the hill of Hermitage, Tain has spruced itself up immeasurably over the ensuing 15 years. Major producers Jaboulet and Ferraton have opened reliable brasseries *(@le.vineum.jaboulet and caveau-ferraton.fr respectively)*. More formal options include the unmissable Le Mangevins *(lemangevins.fr)*, a husband-and-wife team offering precise cooking on a quiet backstreet. Another good option is Marius Bistrot in Chapoutier's hotel Fac & Spera *(facetspera.fr)*, also a good choice for somewhere to stay.

NINE MUST-VISIT WINERIES FOR CAVA

Just outside buzzing Barcelona, Spain's popular sparkling wine Cava awaits discovery. The nearest winery to the city is Alta Alella *(altaalella.wine)*, where cellar tours come with tastings of three or six Cavas. Further afield, Castellroig *(castellroig.com)* has the bragging rights of an 18th-century cellar, converted into a wine museum. Meanwhile, Castillo Perelada *(perelada.com)* has a wine spa, and Codorníu *(codorniu.com)* – where the first Cava was made in 1872 – offers a gran reserva matching and tasting.

Codorníu

Rathfinny

BEST ENGLISH VINEYARDS FOR VISITORS

As wineries have multiplied in England, so have tourism experiences. And two wineries in East Sussex are doing things particularly well. At Rathfinny (rathfinnyestate.com), in Alfriston on the South Downs, owners Mark and Sarah Driver have declared their intention to produce world-class sparkling wines. But they also offer dining options which include fine 'Ottolenghi-style' summer lunches next to the vines, where locally sourced, seasonal ingredients accompany your chosen fizz. Biodynamic winery Tillingham (tillingham.com) also puts food first. Cool 30-somethings come to this mixed farm estate near the characterful fortified hilltop town of Rye to devour wood-fired sourdough pizza in a Dutch barn and enjoy a glass or three of the easygoing, quirky wines.

Santa Sofia

GARDA DOC WINERIES

Lake Garda, with its seductive Mediterranean climate and lush vegetation, soft breezes and striking mountain backdrop, is no stranger to travellers. Or wine lovers: its local Valpolicella and Bardolino are famous worldwide. But there's another player on the wine scene that's making a name for itself. Garda DOC cradles the western, southern and part of the eastern shores of the lake, stretching past Verona into Soave and the Lessini Durello sparkling wine area. Thanks to its fresh acidity and complexity, Garganega has become the DOC's leading grape. Look out for producers such as contemporary Perla del Garda *(perladelgarda.it)*, making exquisite Garda DOC spumante and red wines, or Santa Sofia *(santasofia.com)*, which makes white Garda DOC alongside Valpolicella.

A TASTE OF FRANCIACORTA

Have dinner in any self-respecting restaurant in Italy and you'll be offered a glass of Franciacorta. The sparkling wines are Italy's answer to Champagne: high-quality bubbles made using the traditional method. And given that this region near Milan is compact, just around 25km by 10km, it's perfect for a long weekend. Start at the Berlucchi cellars *(berlucchi.it)*; Guido Berlucchi and his oenologist Franco Ziliani – who died in 2021 at the age of 90 – were the 'grandfathers' of Franciacorta, who began making sparkling wines here more than 60 years ago. Carry on to organic Ca' del Bosco *(cadelbosco.com)*, where a modernist cellar and sculpture park is ideal for art fans. Finally, try Vigneti Cenci *(vigneticenci.com)*, a small, family-run estate in an 18th-century farmhouse on the slopes of Monte Orfano. Its courtyard, with shaded tables and overhanging vine pergola, is the perfect place to taste after a walk in the vineyards.

Ca' del Bosco

Peloponnese vineyards

30

THE PELOPONNESE: WINE DESTINATIONS, AND A WALK

The Peloponnese peninsula, in the heart of the Mediterranean, south from mainland Greece, challenges the easy stereotypes of what Greece offers visitors: warm seas, sun-drenched white terraces, fresh seafood... Yes, there is all of that, but also snow-peaked mountains, endless historical landmarks – and wine. The Peloponnese is home to the largest number of Greek wine PDOs, and its variety of terroirs has made it an area of wine production since ancient times. A compulsory stop is Nemea, an appellation producing intense reds made from the local Agiorgitiko variety. Here, you will find no shortage of wineries organising tastings, including Seméli *(semeliestate.gr)*, Gaia *(gaiawines.gr)* and Lafazanis *(lafazanis.gr)*. Afterwards, the nearby village of Vytina is known for its feta cheese and honey, made by bees in the surrounding hills. From there you can easily reach the nearby Lousios Gorge monastery trail, which is one the most beautiful hikes in Europe.

GEORGIA: TOP DESTINATIONS FOR WINE LOVERS

Georgia's wine heritage is like no other. Its language gave us the very word 'wine' (from 'gvino'), and winemaking has existed here for at least 8,000 years. Georgia also has an impressive number of indigenous grape varieties. Roughly an hour's drive east from capital Tbilisi, Kakheti is its biggest and best-known region, where you will find the most-planted grapes: Saperavi and Rkatsiteli. Here, with its hilltop location overlooking vast Alazani valley, cobbled streets and terracotta-tiled buildings, Sighnaghi is perhaps the quaintest village in Georgia. It is the perfect place from which to explore local wineries, most notably Pheasant's Tears *(from 'Georgianised' American winemaker and artist John Wurdeman: pheasantstears.com)*. Another winery to seek out is Nikalas Marani *(nikalasmarani.ge)*. Owner Zurab Mgvdliashvili produces natural wines from grapes Rkatsiteli, Mtsvane, Kisi, Saperavi and the resurgent Khikhvi. These are wines of rare vitality and character; tasting them is an unforgettable experience.

Nikalas Marani

Bravio delle Botti

MONTEPULCIANO WINE TOUR, TUSCANY

Made an hour from Siena, Vino Nobile di Montepulciano is Brunello's younger sibling – softer, more playful, with a mischievous spring in its step. The crowds flock to Montepulciano town in August for the annual Bravio delle Botti *(braviodellebotti.com)*, a spectacle in which muscular men attempt to push heavy wine barrels up cobblestone streets. But really, it's lovely throughout the year. The way to producer Avignonesi *(avignonesi.it)* is along an avenue of cypress trees, into the heart of the 200ha Le Capezzine estate. A beautiful brick facade beckons you into a modern yet respectfully rustic loft conversion where you can taste wines. Try the silky soft Grandi Annate, Vino Nobile di Montepulciano 2012, all wild roses and plums. And do not even consider leaving without having meditated over the legendary vin santo, a hymn to the heavens.

NAPA VALLEY: WHERE TO VISIT, EAT AND STAY

Here's how to spend a perfect 24 hours in Napa Valley. Follow locals to breakfast at Contimo Provisions *(contimonapa.com)* in downtown Napa. Take its legendary ham & jam biscuit on a stroll along the waterfront, over First Street bridge to Oxbow Public Market, where you can grab a coffee. Book the 12pm slot of Ashes & Diamonds' 'A&D Wines + Food' experience, which covers both wine tasting and lunch *($165 for 60-90 mins: ashesdiamonds.com)*. On the outskirts of town, this stark, mid-20th-century winery looks unlike any other estate in Napa Valley and represents a new era of Napa wines: approachable and lower alcohol. In the afternoon, book a tasting at Faust Haus *(faustwines.com)* in St Helena, north on Highway 29/128. On its patio taste through elegant, cooler-climate Cabernet Sauvignons. Head a little further up Highway 128 to Brasswood Bar & Kitchen *(brasswood.com)* for a pre-dinner snack (the off-menu, hand-pulled Mozzarella al Minuto), then have dinner at Bistro Don Giovanni *(bistrodongiovanni.com)*, next door to Ashes & Diamonds Winery back in Napa. Request an outdoor table by the fountain and order the seared salmon fillet. Nightcaps are in downtown Napa at the women-owned Cadet Wine & Beer Bar *(cadetbar.com)*, stumbling distance from the minimalist Archer Hotel *(archerhotel.com)*.

Brasswood Bar & Kitchen

26 DREAM DESTINATION: ROSEWOOD CASTIGLION DEL BOSCO, TUSCANY

Undulating valleys, flecked by pin-straight pines and flanked by woodlands. Sun-drenched olive trees and slopes carpeted in vines. And, perched at the top of a hill overlooking it all, Rosewood Castiglion del Bosco (rosewoodhotels.com), sitting northwest of Montalcino in the Val d'Orcia nature reserve. Tuscan hotels don't get more picturesque than this 2,000ha estate. Turn the clock back a millennium and this remarkable enclave housed a medieval castle. After falling into disrepair, it was purchased by fashion scion Massimo Ferragamo in 2003 and splendidly regenerated into a resort. He was drawn, as all wine-loving visitors will be, by Castiglion del Bosco's deep connection to Brunello di Montalcino. The estate has grown grapes for centuries, and in 1967 Castiglion del Bosco was one of the founding members of the Consorzio del Brunello di Montalcino. Visit today and you'll see that the connection continues to thrive. Once you've dropped your bags in your suite, you can be at the on-site winery sipping Sangiovese within minutes.

View from the veranda at Osteria La Canonica, Castiglion del Bosco

BEST WINE BARS IN NAPA & SONOMA

In Napa and Sonoma, a wave of new venues has entered the scene – with obscure wines and trendy designs made for the Instagram age. This is good news for visitors, for there is now literally a wine bar for everyone. For example, take the forward-thinking tasting room Region *(drinkregion.com)*, opened in Sebastopol. Wine is dispensed from self-serve stations at the tap of a button and, with 50 local producers on offer, you can sip your way around Sonoma County's AVAs. Most of Region's wine partners don't have their own tasting rooms, so the bar features a different winery each week as a pop-up, offering the chance to meet winemakers and try new releases and library vintages.

25

Region, Sebastopol

MENDOCINO & ANDERSON VALLEY TOUR

Now a destination in its own right, California's cool-climate Anderson Valley has developed into more than just a stopover on the way to the Pacific coast. And idyllic Mendocino makes the perfect base. Founded in 1851, Mendocino's quaint charm makes it feel like a town frozen in time. It wasn't until the 1970s that the first commercial wineries planted vineyards in Anderson Valley. Cool weather made it harder for thicker-skinned red grapes to fully ripen, so the initial focus was on white varieties such as Riesling and Chardonnay. But Pinot Noir became increasingly popular in the 1990s, as the potential for the grape in Anderson Valley was more widely recognised. Today, estates such as Drew Family Wines *(drewwines.com)* turn out award-winning Pinot Noir from fruit grown primarily in fog-shrouded, high-elevation Mendocino Ridge vineyards.'

24

Drew Family Wines

TOP ORGANIC NAPA WINERIES

Organic and sustainable farming practices were part of the ethos in Napa Valley long before they became trendy. Today, many of Napa's organically certified wineries gladly open their cellar doors to provide a more in-depth look at the organics and invite visitors to taste the difference. For example, at Frog's Leap Winery *(frogsleap.com)* in Rutherford, owner-winemaker John Williams is a lifetime disciple of environmentally sound farming – threading together biodynamics, organics and sustainability to ensure vineyard health. Frog's Leap has been solar-powered since 2005, certified organic since 1989 and was Napa Valley's first LEED (Leadership in Energy and Environmental Design) certified winery. Meanwhile, Robert Sinskey Vineyards *(robertsinskey.com)* has organic certification and most of its land is irrigated using reclaimed winemaking production water, treated in engineered wetlands.

Frog's Leap Winery

23

Les Doux Secrets d'Hélène

22

TOP BORDEAUX WINE BARS

Hidden away, steps from Bordeaux's Grosse Cloche belfry, Les Doux Secrets d'Hélène on Rue Neuve is practically a secret itself; this is not a wine bar you'll just stumble upon. With the cosy couches, antique sewing machines repurposed as tables, and a massive fireplace, it's the kind of place you visit for a romantic evening out or a quiet conversation with friends. Former sommelier Hélène Orhon has curated an extensive list mostly from around France, but also featuring wines from Croatia, Italy, Chile and South Africa, too. Orhon is incredibly knowledgeable and it's easy to discover lesser-known bottlings here, with a good selection by the glass. Try the gourmand pairing, which comes with five different decadent bites crafted with local, seasonal products.

SANTORINI: OLD VINES & SPECTACULAR VIEWS

On sun-soaked Santorini in Greece, distinctive wines are produced from resilient grapes. After 100 years of trading from an old canava in Episkopi Gonia, Argyros *(estateargyros.com)* built a modern winery with an architecturally striking visitor centre. Out in the dusty fields there are vines that date back to the 1850s, some of the oldest on the island. The fourth generation of the family now produces Assyrtiko wines including small-batch cuvées, an Aidani, a full-bodied Mavrotragano red, its Atlantis red blend of Mandilaria and Mavrotragano from 'relatively young vines' (60 years or younger), and a selection of vinsantos.

On the island's west is Venetsanos *(venetsanoswinery.com)*, the winery with the most dramatic location on Santorini *(pictured)*, built into a cliff in 1947. In a time before available electricity, engineer Giorgos Venetsanos had the idea of using gravity to process grapes, piping the resulting wine to the harbour 230m below for export. Tastings are held on a rocky terrace with the stunning caldera below.

Venetsanos Winery's terrace

21

Château Le Thil

20

STAY IN A BORDEAUX CHÂTEAU

There are hundreds of accredited châteaux that welcome guests in Bordeaux – and some you can take over exclusively. It's a great opportunity to discover local architecture from the inside, and the invitation can include private access to cellar visits, tastings and seriously good food designed to complement the wines brought up directly from the cellar. Options include Château Le Thil (chateau-le-thil.com), built in the traditional Bordeaux architectural 'chartreuse' style, and set in woodland in Pessac-Léognan. Owned since 2012 by the Cathiard family – who own five-star hotel and spa Les Sources des Caudalies, nearby – Château Le Thil was opened as a 'lodge' in 2013. Stone stairways at opposite ends of the house lead up to nine tastefully decorated rooms and two suites.

La Petite Colombe, Leeu Estates

19

LUXURY TRAVEL, SOUTH AFRICA: TOP WINE HOTELS

From dining next to Table Mountain to whale watching in Walker Bay, South Africa's finest wine hotels come with serious wow factor. In Stellenbosch, Delaire Graff Lodges & Spa *(delaire.co.za)* has jawdropping views and will reopen in October 2023 after an update to the enormous free-standing suites, each with its own plunge pool. The hotel is decked out with hundreds of artworks from owner Laurence Graff's private collection, allowing you a glimpse into contemporary South African art. Generally regarded as the most luxurious and exclusive Winelands hotel, facilities include a private cinema, spa and tasting room.

In Franschhoek, Leeu Estates *(leeucollection.com)* is within walking distance of the Wine Studio, where you can taste wines made by talented duo Chris and Andrea Mullineux. The restored 19th-century Cape Dutch manor house is set in landscaped gardens, and is home to a large contemporary art collection, as well as a pool and spa, and La Petite Colombe, one of the top 10 restaurants in the Cape. A complimentary shuttle will take you to Franschhoek's high street, just a two-minute drive away, for shopping and dining.

Meanwhile, in Constantia, near False Bay, the boutique Steenberg Hotel & Spa *(steenbergfarm.com)* is set in a 17th-century manor house, and in two- and three-bedroom villas. The Steenberg 18-hole golf course, and two destination restaurants, are additional draws.

A PERFECT WEEKEND IN BEAUNE

Hospices de Beaune

Quaint cafés, sun-drenched terraces, wine-soaked meals all day long – what more could you want? The Burgundian city of Beaune is the perfect place for a weekend escape. Kick off with breakfast at boutique hotel L'Imprimerie *(limprimeriebeaune.com)* before hitting the Beaune Saturday market to scour the fresh produce and pungent cheeses. After a picnic with your finds in Parc de la Bouzaize a little out of the centre, stock up on artisan bottles from small producers at Avintures *(avintures.fr)*. Later the Musée du Vin near the cathedral in the centre is the place to brush up on the history and climats of Burgundy, as well as the intricacies of viticulture and vinification. Dinner is at Le Comptoir des Tontons on Rue du Faubourg Madeleine. Easy-to-share tapas are crafted from local and organic products, served alongside a selection of French wines. On Sunday, don't miss a visit to the famed Hospices de Beaune, founded in 1443 and operating as a charity wine auction house since 1859 *(beaune-tourism.com)*.

RIOJA: WINE TOUR HIGHLIGHTS

An old marketing slogan defined Rioja as The Land of a Thousand Wines. Such a claim may sound exaggerated, but it feels less ridiculous once you discover the region's diversity, which goes well beyond its wines. Spanning more than 100km west to east along the Ebro river, the region has seven river valleys, hilltop towns set against mountain ranges, award-winning architecture and ancient monasteries straddling the Camino de Santiago pilgrims' trail. Just north of Laguardia, Bodegas Ysios *(bodegasysios.com)* and its wavy metal roof provide the perfect Instagram-ready shot, especially if you manage to catch the clouds sliding over the crest of the Sierra Cantabria. If you'd prefer to enjoy this view from a chaise longue, glass of wine in hand, try the stylish wine bar outside the gates of Bodegas Javier San Pedro Ortega *(bodegasjaviersanpedro.com)*. Javier, a fifth-generation grower, makes a diverse range of wines, from aromatic whites to serious single-vineyard reds.

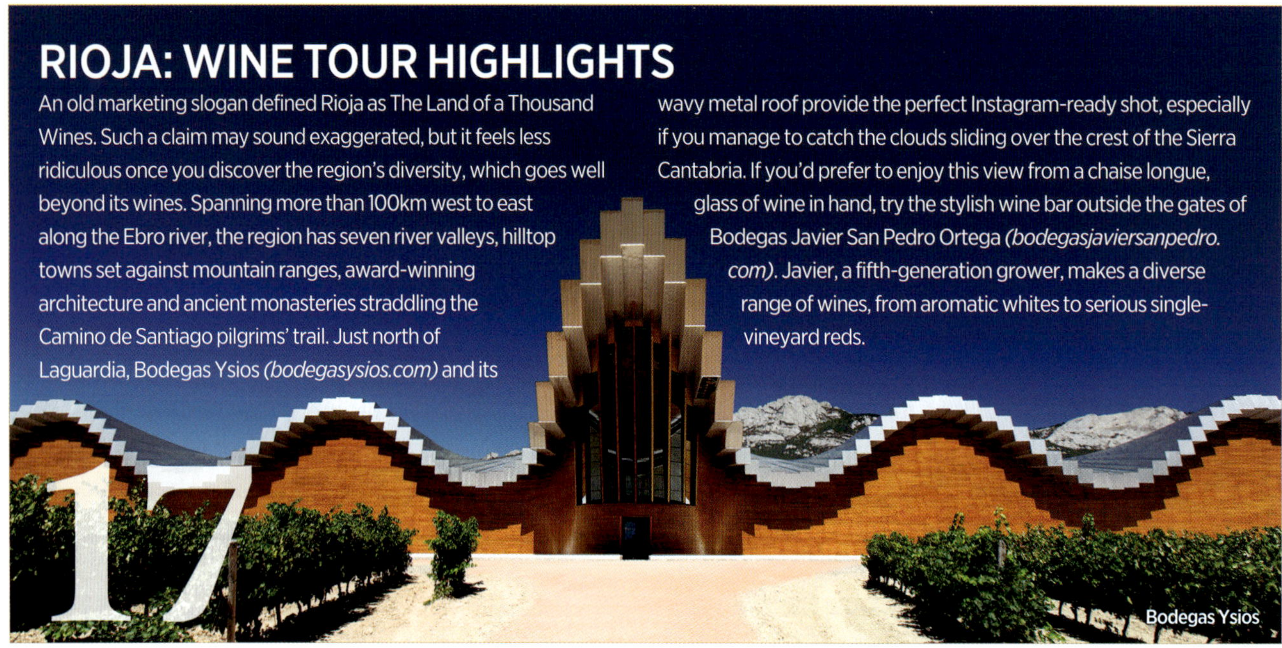

Bodegas Ysios

BEST BORDEAUX HOTELS FOR WINE LOVERS

Accommodation options and activities in and around Bordeaux are constantly expanding – and you'll miss out if you look only in the city centre. Staying in the vineyards is much more authentic. Foodies should head for Château Lafaurie-Peyraguey in Sauternes *(lafauriepeyragueylalique.com)*, for the two-star Michelin restaurant and the wine and Lalique crystal shop. Other foodie options include Château Troplong Mondot's Michelin one-star restarant Les Belles Perdrix in St-Émilion. Or dine at the 'private table' of Château Haut-Bailly near Léognan, for gourmet lunches and dinners for 4-15 people *(via haut-bailly.com/en/hospitality.html)*. Close by, adjacent to Château Smith Haut Lafitte, five-star hotel & spa Les Sources de Caudalie also offers fine dining at Michelin two-star La Grand'Vigne, as well as its more bistro-style La Table du Lavoir restaurant.

History buffs will love Château Pape Clément at Pessac; families are very comfortable at Château Biac in Langoiran; or if you'd like something a little different, you can sleep in a repurposed wine vat at eco-friendly Château Bonhoste in Entre-deux-Mers. With tourism in full swing, there's something for everyone in Bordeaux.

Château Lafaurie-Peyraguey

Villa Medicea di Lilliano

TOP WINERY HOTELS NEAR FLORENCE

Too often overlooked in favour of neighbour Chianti Classico, the Chianti Colli Fiorentini and Chianti Rùfina regions near the Tuscan capital boast centuries-old estates making authentic, approachable wines. The Florence fringe offers dreamy views and country-chic accommodation – plus ancient castles and noble villas. Set among 60ha of organic olive groves and vineyards, the 1,000-year-old wine estate Villa Medicea di Lilliano *(medicivilla.com)* served as a Medici residence. The five individual villas exhibit contemporary Tuscan design with traditional features: terracotta-tiled floors, beamed ceilings, Chianti-coloured cushions and wall art displaying Florentine scenes. Take a tour of the villa, from the terrace and its distant views of Brunelleschi's dome to the sumptuous frescoed rooms, before strolling around historic cellars and sipping a glass of the estate's fruit-forward Chianti Colli Fiorentini or harmonious Canaiolo.

Morgado do Quintão

ALGARVE BEYOND THE BEACHES

When the Romans reached the southwestern point of the Algarve, they thought it was the end of the world, where the waters of the ocean boiled at sunset. Still they planted vines in the region, finding the temperate climate and fertile terroir a nirvana for wine-growing. Fast forward to present times and it isn't just grapes that relish soaking up the rays, as beach lovers bask in the reputed annual 300 days of sunshine.

At first, coastal vineyards lost out to the package holiday seaside resorts developed from the 1960s. But in a land deeply rooted in the traditions of wine-growing, artisanal viticulture is re-emerging in a revival of indigenous grape varieties, especially the revered Negra Mole (meaning 'black soft').

As enterprising estate owners such as Quinta do Barranco Longo *(quintadobarrancolongo.com)* and Morgado do Quintão *(morgadodoquintao.pt)* become increasingly recognised for the quality they are achieving, a new set of adventurous wine tourism thrill-seekers is being drawn to Portugal's south. This final frontier of western Europe still has so many grapes yet to be tasted and explored.

BEST WINERY HOTELS IN SPAIN

'Luxury, thy name is Abadía Retuerta! This 12th-century monastery-fortress (abadia-retuerta.com) in Castilla y León – near Sardón de Duero, just beyond the western extremity of DO Ribera del Duero – may look austere from the outside, but step inside for the utmost spoiling. The first clue is the Michelin one-star Refectorio restaurant, for which you will have wisely booked ahead. The second clue is at check-in when you're given a mobile to call your personal butler. In your room, treats await; maybe a bowl of cherries. Subtle decor offsets the pale stonework – the effect is utterly soothing. The local countryside isn't scenic, so dedicate your hours to the spa, restaurant, pool bar and wines (which are worth tasting). Wander round the cloister and church, and book a trip to explore the impressive estate. You will want to return.'

Abadia Retuerta

Tormaresca

EXPLORING PUGLIA

For many years certain English people spoke of Puglia – the long heel of the Italian boot – as their new Tuscany, ripe for discovery. Funnily enough, for centuries Puglia's thriving wine industry has out-produced Tuscany in terms of volumes. Historically, however, most of Puglia's grapes were sent to create vermouths in Piedmont or blended with other wines made by large-scale urban producers. It was only late in the 20th century that local producers began to bottle and market their wines as Puglian.

This growing movement was encouraged when celebrated producer Marchesi Antinori *(antinori.it)* began investing in the region from 1998, going on to open two wineries at opposite ends of Puglia to do for this region what it had done for Tuscany in the 1970s. West of Bari, near the border with Basilicata, Tormaresca's *(tormaresca.it)* Bocca di Lupo vineyard has an extensive, modern winery-cum-visitor centre inspired by typical Puglian fortified masseria buildings. Wines produced here include Bocca di Lupo, a deep ruby Aglianico aged for 15 months in French oak (2018, £66 Hic, Strictly Wine). The building is calm and spacious, and exudes the self-confidence of the Antinori operation.'

BURGUNDY BY BIKE

You can walk out to the vineyards from Beaune centre in about 20 minutes. But hiring a bike is a great way to see more of the Burgundian vines if you have time. Bike hire is available at Bourgogne Evasion (bourgogne-evasion.fr), on the edge of the Parc de la Bouzaize, allée du Dr Bouley, as you head out of town for the Voie des Vignes route. This designated cycle track runs right through the vineyards and is predominantly flat for the 20-or-so minutes it takes to reach Pommard to the southwest. There is then a mild climb up to Volnay, and you can even carry on all the way down the Côte de Beaune, through Meursault, and eventually – if you're feeling the need to work off last night's dinner – ending around Santenay.

TOP CHAMPAGNE HOTELS OFF THE BEATEN TRACK

Wine enthusiasts exploring Champagne face a challenge. Visiting the famous houses' cellars and reception centres in Reims and Epernay often means missing out on the vineyards, scenery and myriad small family producers of the region – which are all situated outside the main towns. But rural hotel havens get you up close to the soul of the world's favourite fizz. For example, Clos 1753 in Ecueil may be the best gite-type self-catering hideaway in all Champagne. The property is situated in the courtyard at Champagne Nicolas Maillart *(champagne-maillart.fr)*, and offers four bedrooms and oodles of living space; tours and tastings need to be booked ahead. If the tranquillity begins to irk, Reims is only 20 minutes away by car, or you could visit other nearby star producers Champagnes Savart *(champagne-savart.com)* and Lacourte-Godbillon *(champagne-lacourte- godbillon.com)*.

Set in the west of the Vallée de la Marne region, Auberge le Relais *(relaisreuilly.com)* at Reuilly-Sauvigny is perfect for visits to elite small growers and the sleepy villages along the winding Marne river. Expect smiling and prompt service, and great vistas over the valley from the south-facing rooms – as well as a rather good restaurant where the wine list is gold dust (with bottles at prices which would make Paris blush, too). Nearby, at Fossoy, visit Champagne Benoît Déhu, producer of La Rue des Noyers, one of Champagne's greatest single-vineyard wines.

Finally, the Aube is Champagne's current hotbed of creativity and informed Champagnistas will want to spend time there. There is now good accommodation in the magical countryside, saving a stay in Troyes, an hour's drive from the Côte des Bar vineyards. The River House is a tiny Seine-banked hideaway in the western Aube – the Barséquanais area *(legardechampetre.fr)*. The rooms have a farmhouse-meets-cool boutique hotel vibe. Make time to visit nearby Champagnes Cedric Bouchard, Devaux *(champagne-devaux.com)*, Domaine La Borderie *(champagne-domaine-la-borderie.fr)*, biodynamic Marie Courtin and Vouette et Sorbée *(vouette-et-sorbee.com)*.

Auberge le Relais

Champagne Nicolas Maillart

Sierra de Tramuntana

9

MALLORCA'S RENASCENT WINE SCENE

Balearic island Mallorca has a reputation as a beachy, sunseeker hotspot. But most of the island north and east of capital Palma is tranquil and bucolic, with charming, uncrowded rural landscapes, hamlets and villages. Explore the Tramuntana mountain range, where you can go climbing or hiking *(mallorca.com)*; and cycle between wineries across the plain of Es Pla, with its low stone walls and peach and almond orchards.

Viticulture and winemaking on the island date back at least to the Romans, but in the 19th century the vineyards were devastated by phylloxera. Viticulture didn't really recover until the late 20th century, when tourist demand for local wines led to the creation of the DOs of Binissalem and Pla i Llevant. This has led to Mallorca having a surprising number of small wineries – approaching 100, estimates say – geared towards quality wines and wine tourism.

Located in the centre of Manacor on the east side of the island, family-run Vins Miquel Gelabert *(vinsmiquelgelabert.com)* was founded in 1985, when winemaking in Mallorca was just beginning its recovery. Ex-chef Miquel Gelabert is affectionately known as 'the madman of Manacor' due to the sheer number (up to 27) of individual wines he produces annually, along with his oenologist son. They also cultivate more than 30 native and international grape varieties. Not-to-be-missed classics are Gran Vinya Son Caules and Sa Vall Selecció Privada.

Half an hour's drive from Palma, in Algaída, Can Majoral *(canmajoral.com)* was founded in 1979 by Andreu Oliver and was the second producer in Spain to achieve organic certification. Oliver and his team have added a range of native grapes – including Callet, Giró Ros, Gorgollassa and Manto Negro – to the original international varieties to create characterful quality wines under the Pla i Llevant DO.

SOUTHERN ITALY: WINERIES & VINEYARD STAYS

The Tasca d'Almerita family has long been considered the royalty of Sicilian winemaking. Its headquarters are in the Sicilian heartlands at Regaleali, but in recent years its estates have expanded into other parts of Sicily. The jewel is Capofaro Locanda & Malvasia *(capofaro.it)* on the island of Salina, one of the volcanic Aeolian islands that belong to Sicily.

Capofaro is the perfect idyllic getaway for wine lovers. The 27 rooms are set among vineyards where the grapes for the delicious dessert wine Malvasia delle Lipari are grown. The estate overlooks the sea, with beaches nearby, plus a central pool at the resort itself. The restaurant offers the best of the Mediterranean: fresh seafood, sun-nourished vegetables such as olives and wild herbs. Recipes are created from the many cultural influences that form Sicily's well-flavoured cuisine, including rustic peasant dishes and aristocratic food from the region's golden age.

Planeta *(planetaestate.it)* is another major name in Sicily, the first winery with a vision to communicate the island's viticultural greatness to a modern international audience. The Planeta family has always understood the value of Sicily's diversity and has been enthusiastic in helping to build wine tourism on the island through its hospitality. Its headquarters are in Menfi, on the southwest coast of Sicily, and that's where the Planetas have also created their country house hotel. La Foresteria offers 14 rooms, a stunning infinity pool, herb gardens and beach access. A relaxed, country-chic aesthetic runs through the bedrooms, the large kitchen and reception rooms. There's great food to be had, with cooking classes – as well as wine tastings from all the family's estates. In warm weather, eat outside on the terrace overlooking the vineyards after an afternoon trip to the Greek-style temples of Selinunte and Segesta.

Capofaro Locanda & Malvasia

MONTALCINO: ENJOY FOOD & WINE IN BRUNELLO COUNTRY

Castello Banfi's La Taverna restaurant at Poggio alle Mura

Montalcino is a superstar wine region, fortuitously off the beaten track. Unspoiled, with no motorway nearby, the most ubiquitous through-traffic is human-powered. Steady streams of pilgrims plod the ancient Francigena byway crossing Montalcino, heading south to Rome's Vatican. Packs of Lycra-clad cyclists pedal themselves to exhaustion along the bumpy chalk byways of the Eroica (or 'Heroic') Route. And wine lovers come to taste the world's most famed 100% Sangiovese red wines: the oak-aged Brunello di Montalcino DOCG, and its earlier-bottled, no-oak-needed sibling Rosso di Montalcino DOC.

Montalcino's name derives from Monte Leccio ('holm oak hill'). Evergreen oak forests cover more land here now than they did in 1860. They host roebuck, edible mushrooms, wild asparagus, truffles and wild boar: fare integral to a seasonal local food culture. Montalcino also produces renowned honey, with beekeepers from across Italy renting space here for their hives. Shielded by Monte Amiata, central Italy's highest peak, Montalcino is a warm, breezy spot that's perfect for Brunello vines. Producers to visit in the area include large-scale Castello Banfi *(banfi.it)*, located in Montalcino's far southwest, and Col d'Orcia *(coldorcia.com)*, maker of musky Moscadello (a sweet white wine which pre-dates Montalcino's Brunello reds by centuries).

If you want to compare the differences between Brunellos from two different sides of the same hill but under the same owner, visit Elisabetta Gnudi Angelini's Altesino and/or Caparzo wineries.

The area's outstanding natural beauty is matched by its timeless architecture, from breathtaking fortresses and walled castles to small farmsteads with open-hearth fires. Aesthetically, all must (by law) respect tradition, even down to the exact colour and shape of window shutters and roof tiles. Staying in these historic places is part of the regional experience. A classic, fairly priced overnight option on the edge of Montalcino town is Hotel Vecchia Oliviera *(vecchiaoliviera.com)*, a former frantoio olive oil mill.

CALIFORNIA: THREE DAYS IN SONOMA

Despite living in its neighbour's shadow, Sonoma wine country is double the size of Napa Valley, extending to destinations such as Healdsburg and the Russian River Valley, the funky town of Sebastopol and even the Pacific ocean. And while Napa has zeroed in on Cabernet Sauvignon and Bordeaux varieties, Sonoma is far more diverse. It's home to more than 60 grape varieties (though Pinot Noir and Chardonnay are the stars) and 400 or more wineries spread over (now) 19 AVAs.

Due to its vastness, there are two common mistakes visitors to Sonoma make. They either plan only a day or two, hardly scratching the surface, or they schedule winery appointments, meals and hotel stays without realising it can easily take an hour to get from one to the next. So, good planning is key. And whatever your chosen itinerary, one essential stop is Sonoma town. A one-hour drive from San Francisco, it is anchored by the charming and historic Sonoma Plaza. The birthplace of the California flag, this square played a pivotal role in the state's declaration of independence from Mexican rule. The historic buildings and adobes are now occupied by shops, wine tasting rooms, hotels and restaurants.

Start by sipping Pinot Noir and Chardonnay at Three Sticks Wines *(threestickswines.com)*, housed in the Vallejo-Casteñada Adobe, the longest-occupied residence from California's Mexican Period. Schedule a second wine tasting in the intimate salon of nearby Sojourn Cellars *(sojourncellars.com)*, which sources fruit for its Pinot Noirs and Chardonnays from some of the top vineyards in Sonoma County.

Then check into the MacArthur Place Hotel and Spa *(macarthurplace.com)*. Located within a 20-minute walk of the Sonoma Plaza, it's the place to relax at the pool or spa before you dine at Layla, the hotel's Mediterranean restaurant. After dinner, enjoy a nightcap at The Bar, a Gatsby-era lounge serving signature cocktails and late-night nibbles.

Three Sticks Wines, Sonoma

Cantina Ricchi

NORTHERN ITALY: LAKE GARDA'S VARIED WINE STYLES

Writers have sung praises of its beauty for centuries, including DH Lawrence, who shared his encounters in Twilight in Italy: "The lake lies dim and milky, the mountains are dark blue at the back, while over them the sky gushes and glistens with light."

Lake Garda's charming character is defined by its medieval castles, quaint lakeside villages and crystal blue waters. Surrounded by three regions – Veneto's rolling hills, Trentino-Alto Adige's alpine mountains and the fertile plains of Lombardy – this is also the northernmost Mediterranean climate in Europe, where the sun-soaked area is tempered by predictable afternoon breezes. The landscape is verdant with lemon and olive groves, and vineyards that provide quaffable enjoyment made from grapes that are grown on morainic soils left by glaciers millions of years ago.

Lake Garda vineyards offer some of the country's most classic wines – Bardolino, the area's famous light red wine, as well as refreshing Chiaretto rosatos. Lugana DOC straddles both Lombardy and Veneto on the southern shore of the lake, producing wines made mostly with Turbiana (Trebbiano di Lugana).

The easy-drinking white Custoza is typically blended from nine permitted grapes, among them Garganega, Fernanda (Cortese) and Trebbiano di Toscano. Azienda Agricola Cavalchina (*tenutedifamiglia.com*) pioneered the use of the name 'Custoza' in the early 1960s and you can visit its second-storey tasting room overlooking the picturesque vineyards where Garganega, Fernanda and other varieties grow.

International varieties such as Merlot and Cabernet Sauvignon are also found along the lake region as they have taken fertile residence for centuries. Cantina Ricchi (*cantinaricchi.it*) is one producer growing these grapes, as well as Chardonnay and Cabernet Franc.

Garda DOC is a still-relatively new term referring to sparkling wines of Lake Garda. They are made by either metodo classico and/or Martinotti methods, offering refreshing, clean bubbles that pair effortlessly with the lake's conviviality. Try Azienda Agricola Pratello (*pratello.com*), an 1860s farm and agriturismo run by the Bertola family, for a range including metodo ancestrale, brut, rosé and long lees-matured metodo classico.

Bardolino

BORDEAUX: TRAVEL SECRETS OF THE LOCAL EXPERTS

Château de la Dauphine, Fronsac

The 'terrasse fleurie' tasting area at Château Angludet

We asked Bordeaux's top English-speaking wine-country trip planners and guides to reveal their favourite places to visit. Bordeaux's big-name wineries have traditionally focused primarily on making legendary wines, but an increasing number of properties of all kinds have developed their tourism infrastructure in recent years.

Wendy Narby of Insider Tasting recommends that beginners head to the Entre-deux-Mers region, including a couple of her favourite estates, Château Lestrille *(lestrille.com)* and Château Thieuley *(thieuley.com)*. 'Entre-deux-Mers properties usually produce red, white, rosé, clairet and often sparkling crémant de Bordeaux,' she says, 'so they provide a great chance to taste the whole range.'

Meanwhile, Sarah Seguret of My Bordeaux Tours takes her beginners to Château de la Dauphine in Fronsac *(chateau-dauphine.com)*, where 'a lot of thought has gone into the visits, and the guides are knowledgeable and personable'. St-Émilion itself can get busy, Seguret says; if it is, then this biodynamic vineyard – with its historic house and lovely garden – is the place to turn.

The legendary finesse of Bordeaux wines stems from the blending of various varieties in both white and red wines. For the wine aficionado, a blending class is an opportunity to learn more about this technique. Our experts send their wine-anorak clients to the blending and expert tasting experiences offered by Margaux's Château La Tour de Bessan *(marielaurelurton.com)*, on the Left Bank and, on the Right Bank, St-Émilion's Château de Ferrand *(chateaudeferrand.com)*.

For those travelling with their families, some châteaux also provide entertainment to keep the children occupied while their adults sip the estate wines. Reserve in advance and Château Marquis de Terme *(chateau-marquis-de-terme.com)*, also in Margaux, will mind your children in a playroom, lead them on a sensory tour of the property and provide specially designed workshops, says Narby. Château de Ferrand – owned by Pauline Bich of the Bic pen family – understandably features the best colouring kit for children, according to Martin Hurst of The Bordeaux Concierge, who also notes that this beautiful property has a casual, 'Napa-like vibe', stemming from a renovation that 'respected the grounds and heritage but added unique design elements such as the spectacular revolving tasting bar'. Also in St-Émilion, young ones delight in the 'little experts tour' at Château La Dominique, says Caroline Matthews of Uncorked Wine Tours. [Contact estates directly to confirm current schedules of tastings, tours and children's activities.]

And for history buffs? The spectacular Château La Tour Carnet *(bernard-magrez.com)* at St-Laurent-Médoc, with its stone bridges spanning the medieval castle moat, is living history. Inside, there's a jaw-dropping assortment of antique treasures – from coats of armour to historic harpsichords – that fill the stately panelled rooms. A 1.5-hour wine tour and tasting (€20 per person) covers seven centuries of local history, culminating in a taste of three of the property's wines, with other more in-depth sessions also available.

Nicolle Croft from *wineguidebordeaux.com* recommends soaking up the history at Château Angludet *(chateau-angludet.fr)* in Margaux, owned by the Sichel family, who have been in the Bordeaux wine trade and viticulture for six generations; daughter Daisy Sichel leads the wine education program, with a range of well-priced tastings and experiences.

Another Margaux property, Château Prieuré-Lichine *(prieure-lichine.fr)*, offers a historical overview tour (from €18-€30 per person, according to group size) covering the Médoc that culminates in a tasting.

3

Opus One, Oakville

NAPA & SONOMA TOP WINERIES FOR VISITOR TASTINGS

Napa has 16 appellations – known as American Viticultural Areas (AVAs) – spread through the mountains, valley floor and outlying areas, all with a vinous voice of their own as elevation and terroirs change. There are two main roads that cut through it all, running north to south. Highway 29, where many high-profile wineries are located, is the most frequented, flanked to the east by rolling hills at the base of the Vaca range. The Mayacamas range on the valley's western side separates Napa from the cooler, marine influences experienced in neighbouring Sonoma Valley.

Among the big Highway 29 names is Opus One in Oakville *(opusonewinery.com)*, the project between Baron Philippe de Rothschild of Château Mouton Rothschild and Robert Mondavi. Started in the 1970s and now one of Napa Valley's iconic names, it's a luxurious place to taste a real collector's wine. The basic Courtyard experience starts at $100, but for truly breathtaking views, opt for the Partner's Room tasting experience ($200).

Meanwhile, located off Napa's other main road – the quieter Silverado Trail – Stag's Leap Wine Cellars (stagsleapwinecellars.com) is one of the estates that put California wine on the map in the acclaimed Judgement of Paris tasting in 1976. The Fay visitor centre has stunning views of the vineyards and Stags Leap District AVA and you can try the winery's three top bottlings: Fay, SLV and Cask 23. But not all tastings will feature these wines, so check the descriptions. If you have deep pockets, indulge in a full culinary experience with a seasonal four-course tasting menu.

To the west, Sonoma Valley is larger and more spread out than Napa Valley. With [now] 19 AVAs from diverse terroirs, it's a fascinating place for wine tourism. A fantastic base to explore the region is the charming town of Healdsburg, situated in Russian River Valley AVA. Here at Ridge Vineyards, Eric Baugher has in recent years taken over from the retired Paul Draper – a giant of the California wine industry – as winemaker for Monte Bello Cabernet, which is from a site south of San Francisco. Baugher and his winemaking colleague John Olney continue Draper's legacy of producing some of the best Zinfandels in Sonoma. You can find the Ridge tasting room overlooking the Lytton Springs vineyard, just 15 minutes' drive north from Healdsburg. One of the best-value tastings in the area, flights will include the Estate Chardonnay, different site-specific Zins, perhaps a Mataro (Mourvèdre) and even a precious drop of Monte Bello.

Also just above Healdsburg, small family-run farm Unti *(untivineyards.com)* is influenced by the wines of Tuscany and Campania. Tastings are held in a barn, where GSM blends and Zinfandels sit alongside juicy Sangioveses and Aglianco. Likewise, Idlewild Wines *(idlewildwines.com)* nearby focuses on varieties from Italy's Piedmont and northwest. Enjoy a plate of antipasti with your tasting flight, and expect detailed commentary on the wines themselves.

SEVEN OF THE BEST CHAMPAGNE HOUSES FOR VISITORS

A UNESCO World Heritage site just a 40-minute train ride from Paris, the Champagne region is understandably a hit with visitors. With so many houses to choose from you won't be able to tick off all your favourites on just a short visit, so you need to prioritise. Here are a few with excellent visitor experiences – all in the easy-to-reach city of Reims.

Founded in 1768, Ruinart *(ruinart.com)* is the longest-established Champagne house of all, claiming prime position at the top of the hill on Rue des Crayères and possessing the deepest and most spectacular crayères (chalk cellars) in the region – the only ones to be classified a national monument. Ruinart was the first to use these enormous chalk-mine caverns to age its Champagnes, and the score-marks of their third-century Roman creators are still visible. A visit to the region is incomplete without an opportunity to explore this underworld, topped off with a tasting of the graceful, Chardonnay-focused cuvées of this under-recognised house.

Also in the centre of Reims, the house of Veuve Clicquot *(veuveclicquot.com)* is a juxtaposition of old and new, with a reception room in the trademark Clicquot mango-orange livery perched directly atop its splendid crayères. A dizzying annual production of up to 20 million bottles makes this the second-largest Champagne house of all, with a compelling focus on the power of the Pinot Noir grape in its blends. A guided cellar tour and tasting are a must (starting at €35), and you can also book a 'Madame Clicquot, the Grande Dame of Champagne' tour and tasting for €100.

The caves at Ruinart

However, to truly experience the diversity of the Champagne region, visiting independent growers is highly recommended. Pierre Gimonnet & Fils *(champagne-gimonnet.com)* is one of the largest of these independent growers, providing a good introduction to the north of the Côte des Blancs sub-region, home of some of Champagne's greatest Chardonnay. These are impeccably crisp, pure wines which present a vivid picture of the chalky slopes between Cuis and Oger. Despite being one of the few growers with published opening hours, it is recommended that you make contact to book in advance *(+33 [0]3 26 59 78 70)*.

If accessibility is important, Mumm *(mumm.com)*, in Reims, provides a wonderful introduction to Champagne, taking in its extensive cellars but also its quirkier projects (such as the first bottle of wine in space). During summer weekends, Mumm offers experiences that include a thought-provoking sensory tasting, and Champagne and cheese pairing.

Finally, the Champagnes of the famous family house of Taittinger *(taittinger.com)* are more refined than ever – and soon the visitor experience will match, as it fully reopens in 2024 after renovations. A visit to its 4km of epic crayères is eclipsed only by a custom tasting with a host who'll tell the story of each cuvée. The legendary Comtes de Champagne ranks among the finest blanc de blancs in the world.

Stained glass artwork in the Taittinger cellars, Reims

TUSCANY: THE WINERIES YOU SHOULDN'T MISS

Tuscany is like a great bottle of wine; lovingly created and carefully aged. The longer you spend contemplating its rich hues, the better it gets. The region's undulating hills are home to some of Italy's best-known appellations, including Brunello di Montalcino, Chianti Classico and Vino Nobile di Montepulciano, as well as Vernaccia di san Gimignano and Carmignano. In the coastal vineyards, and particularly by the 'SuperTuscan' wineries around Bolgheri, you can drive along streets lined with 400-year-old cypress trees.

A Tuscan wine tour is not a five-minute tasting at a roadside wine bar. It's an immersive experience into the region's fabric and history, from the ruins of medieval churches to landscapes that inspired Leonardo da Vinci. Food, naturally, is a big part of the wine tasting culture here, so expect to do some first-hand research on Tuscan wine and food pairings.

To begin, what could be more Tuscan than tasting wine in a 1,000-year-old castle owned by Florentine nobility? Great artists such as Donatello and Michelozzo Michelozzi regularly purchased wine from the estate Castello di Nipozzano *(frescobaldi.com)*, built to guard Florence. The property was destroyed in 1944 but has been partially rebuilt, and you can still view the original cellar at the Renaissance villa. It's a true working farm, and a big one at that, with more than 600ha under vine; there are olive trees and an on-site olive press, and Chianina cows can be seen roaming free in the fields. A visit will include a tour of the monumental cellars used to age Chianti Rùfina, the higher-altitude appellation in the Chianti area. Other highlights include the tasting room in an old kitchen, as well as the views of perfectly maintained vineyards.

Meanwhile, about 15km west of Florence, Capezzana *(capezzana.it)* has been producing wine and extra-virgin olive oil since 804. The Contini Bonacossi family has been running the estate since the 1920s and the youngest members (Oscar, Ettore, Giulia and Duccio) created the newest aspect of the winery – a wine bar called La Vinsantaia where guests can enjoy informal wine tastings, as well as food (*book ahead: +39 334 949 9402*). This is a large, diversified estate, with 650ha of forest, organic vineyards and olive groves, plus a cookery school. In summer, don't miss the terrace with the view of Florence's Duomo, and leave room to taste one of Tuscany's greatest wines – vin santo, a dessert wine made from grapes left to turn to raisins on drying racks.

Over towards the coast above Castagneto Carducci, Le Macchiole *(lemacchiole.it)* – a polished, family-run Bolgheri winery – puts the focus on international, single-varietal wines. Bolgheri has become world-famous for its SuperTuscan wines. Run by Cinzia Merli and her two sons, Elia and Mattia Le Macchiole, this winery is less formal than the imposing gate suggests. As you take the small, private tour around the organic Cabernet Sauvignon, Cabernet Franc, Syrah and Merlot vineyards (book ahead via the website: tours start at €50 per person), you'll notice a large mural by Italian street artist Ozmo – similar in style to Banksy – which hints at the family's creativity. A tour of the cellar is followed by a tasting of the best single-varietal Cabernet Francs Italy produces, Paleo Rosso.

Find the full articles in their original context at decanter.com/top-50-travel-2023

Castello di Nipozzano, east of Florence near Pontassieve

Capezzana barrel cellar

PICTURE CREDITS

The publishers would like to thank the following sources for their kind permission to reproduce the pictures in this book.

pp6–7 Getty Images; pp8–9 Chris Down/Wiki Commons; pp12–13 Southtownboy/Getty Images; p14(T) Leonid Andronov/Alamy Stock Photo; p14(B) Hemis/Alamy Stock Photo; p21(T) François Poincet; p21(B) Alain Benoit; p24(T) Guillard Jacques-Scope-Image/Alamy, Centre François Mauriac Malagar; pp26–27 Jaubert French Collection/Alamy Stock Photo; p28(T) Quentin Petit; p29(B) Theo Schuman; p30 Aurélien Terrade; p31(B) Michael Harvey; pp32–33 Hemis/Alamy Stock Photo; p35 Jacques Lange (Paris)/Getty Images; pp38–39 Francesco Riccardo Jacomino/Getty Images; p41(T) Westend61/Lisa und Wilfried Bahnmuller/Getty Images; p43(T) Magali Cancel; p44 Stevens Fremont/Getty Images; p46 Domaine de la Citadelle; www.senanque.fr, iStock/Getty Images Plus; p48 Herve Fabre Photography; p49(T) Les Bories; Niels van Kampenhout/Alamy Stock Photo, www.hotellesbories.com; pp50–51 Xantana/Getty Images, iStock/Getty Images Plus; p53 iStock/Getty Images Plus; pp52–53 iStock/Getty Images Plus; p57(T) ADT Touraine/JC Coutand; p57(M) Eckhard Supp/Alamy Stock Photo; p57(BL) F Godard/Andia/Universal Images Group via Getty Images; p59(T&BL) Eric Sander, Imagenavi/Getty Images; pp60–61 Wolfgang Gafriller; p63 Andreas Tauber; pp64–65 Tobias Kaser Photography; p65(T) Alamy Stock Photo; p66 Gianni Bodini; p70 istock/getty images plus; p71(L) www.artedolcelyceum.it and gazing and grazing; p72–73 Andrew Mayovskyy/Alamy Stock Photo; p74 carla capalbo; p75(T) carla capalbo; p76(L) Michelemondini.com; p76(R) carla capalbo; p77(T) Luca Antonio Lorenzelli/Alamy Stock Photo; p77(B) carla capalbo; pp78–79 Matteo Musetti/RealyEasyStar/Alamy Stock Photo; p80 Tommaso Di Girolamo/AGF/Universal Images Group via Getty Images; p81(T) Rostislav Glinsky/Alamy Stock Photo; p81(B) Luca Appiotti; p84 (L) Alessandro Vecchi; pp86–87 Matteo Carassale; pp88–89 Giuseppe Manzi Photography; p89(R) Marcello Serra; p90(B) Matteo Carassale; p90 Ezio Pietro, Maria D'Onghia; p91(T) Laurent Dupont; p92 iStock/Getty Images; p94 Image Professionals gmbh/Alamy Stock Photo; pp96–97 istock/Getty Images Plus; pp98–99 Fernando Trabanco Fotograffia/Getty Images; pp100–01 istock/Getty Images Plus; p102 efesenko/Alamy Stock Photo; p103 SBMR/Alamy Stock Photo; p106(B) www.guiarepsol.com, p107 www.tabernapalocortado.es; www.dondeviajamos.com; pp108–09 MediaProduction/Getty Images; p110(T) agefotostock/Alamy Stock Photo; p110(B) Design Pics Inc/Alamy Stock Photo; p111(MR) Photononstop/Alamy Stock Photo; 112(L) Gonzalo Azumendi/Getty Images; p112(R) James Sturcke/Alamy Stock Photo; pp114–15 Juergen Schonnop/Alamy Stock Photo; p119 Cosmin Iftode/Alamy Stock Photo; pp120–21 Keith Sutherland/Getty Images; p122 iStock/Getty Images; pp126–27 Jonathan Boncek; pp128–29 Gavin Zeigler/Alamy Stock Photo; p130(B) Carl Timpone; p131 Carl Timpone; p133(T) Len Holsborg/Alamy Stock Photo; p133(BR) Raymond Patrick; pp134–35 George Rose/Getty Images; p136(L) All Canada Photos/Alamy Stock Photo; 137 Brendan McGuigan; p139 Will Paterson Photography; p140(T) Ross Becker/inkling illustration; p140(BR) Kimberley Teske Fetrow; p143 John Coletti/Getty Images; p146(B) Sam Dean; p147(T&B) Sam Dean; p148 Tony Demin; p149(B) Andy Wakeman; pp50–51 Robert Holmes/Getty Images; p152 Stephanie Forrer; p153 Andrea Johnson Photograph; pp156–57 Paul Biris/Moment/Getty Images; p159 Keith Sutherland/Getty Images; p160(T) Dave Hutchison Photograph/Getty Images; p161(T) Performance Image/Alamy Stock Photo; pp162–62 Verity E. Milligan/Getty Images; p166 www.margaretriver.com; pp170–71 Verity E. Milligan/Getty Images; p174(T) Lisa Duncan Photography; p175(T) Richard Briggs; pp176–77 iStock/Getty Images; p178 Design Pics Inc/Alamy Stock Photo; p179(L) Andres A Ruffo/Getty Images Plus; p179(R); p180 iStock/Getty Images Plus; p181(T) Guillermo Mansilla/EyeEm/Getty Images; p182 Guido Aguero/500px/Getty Images; pp188–89 ElOjoTorpe/Getty Images; p190 ElOjoTorpe/Getty Images; p191(T) www.bodegabouza.com; p191(BL) www.deluccawines.com; p191(BM) Per Karlsson – BKWine.com/Alamy Stock Photo; p191(BR) Joerg Boethling; p192 Per Karlsson – BKWine.com/Alamy Stock Photo; p193(T) www.alacarta.com.uy; pp194–95 Fabrizio Troiani/Alamy Stock Photo; p197 Carla Capalbo; p198(T) Carla Capalbo; p198(BL) Kostyantyn Manzhura/Alamy Stock Photo; p199 Aaron Geddes/Getty Images; p199(B); Grethe Ulgjell/Alamy Stock Photo; p202(T) Paul Lawrenson (Kent)/Alamy Stock Photo; p202(M) Francesco Riccardo Iacomino/Moment/Getty Images; p203(M) iStock/Getty Images; p204(BR) Robert Fried/Alamy Stock Photo; p205(TL) SvetlanaSF/Shutterstock; p205(TR) Severin Nowacki; p205(B) Herbert Lehmann; p206(TR) Jose Lucas/Alamy Stock Photo; p207(TL) Massimo Borchi/Atlantide Phototravel/Getty Images; p208(M) iStock/Getty Images; p209(T) Associated Press/Alamy Stock Photo; p210 Davide Lovatti; p211(B) Emma K Morris; p213(B) agefotostock/Alamy Stock Photo; p214(T) Tim Graham/Alamy Stock Photo; p215(B) Blanca Saenz de Castillo/Alamy Stock Photo; p217(T) Somatuscani/Getty Images; p217(B) Matteo Carassale; p218(T) Martino Dini; p219(B) iStock/Getty Images Plus; p222(B) Sylvain Sonnet/Getty Images; p233 (main image) Mick Rock/Cephas Picture Library

Maps: Maggie Nelson

Every effort has been made to acknowledge correctly and contact the source and/or copyright holder of each picture and Sona Books apologises for any unintentional errors or omissions, which will be corrected in future editions of the book.